ADAPTIVE CONTROL PROCESSES:
A GUIDED TOUR

ADAPTIVE CONTROL PROCESSES: A GUIDED TOUR

BY

RICHARD BELLMAN

THE RAND CORPORATION

1961

PRINCETON UNIVERSITY PRESS

PRINCETON, NEW JERSEY

Printed in the United States of America

To

A. Lyapunov, H. Poincaré, and S. Lefschetz

and all other scientists
striving for the betterment
of the human race.

PREFACE

This study is an edited and expanded version of a series of lectures delivered at the Hughes Aircraft Company on invitation from Dr. J. M. Richardson of the Research Laboratories.

The interest in these lectures evidenced by the audience, and by a larger scientific audience around the country, was such that it was easy to accede to his further suggestion that a book be written based upon these lectures.

I wish, then, to thank him not only for the opportunity to deliver these lectures, a stimulating experience in itself, but also for the constant encouragement he has furnished my research in these areas, and for the many valuable discussions that I have had with him concerning the theory and application.

This study was completed as part of the research program undertaken for the U.S. Air Force by The RAND Corporation.

Richard Bellman

March, 1959.

CONTENTS

CHAPTER III

MULTISTAGE DECISION PROCESSES AND
DYNAMIC PROGRAMMING

CHAPTER IV

DYNAMIC PROGRAMMING AND THE
CALCULUS OF VARIATIONS

CHAPTER V

COMPUTATIONAL ASPECTS OF DYNAMIC PROGRAMMING

CHAPTER VI

THE LAGRANGE MULTIPLIER

CHAPTER VII

TWO-POINT BOUNDARY VALUE PROBLEMS

CHAPTER VIII

SEQUENTIAL MACHINES AND THE SYNTHESIS
OF LOGICAL SYSTEMS

CHAPTER IX

UNCERTAINTY AND RANDOM PROCESSES

CHAPTER X

STOCHASTIC CONTROL PROCESSES

CHAPTER XI

MARKOVIAN DECISION PROCESSES

CHAPTER XII

QUASILINEARIZATION

CHAPTER XIII

STOCHASTIC LEARNING MODELS

CHAPTER XIV

THE THEORY OF GAMES AND PURSUIT PROCESSES

CHAPTER XV
ADAPTIVE PROCESSES

CHAPTER XVI
ADAPTIVE CONTROL PROCESSES

CHAPTER XVII
SOME ASPECTS OF COMMUNICATION THEORY

CHAPTER XVIII
SUCCESSIVE APPROXIMATION

ADAPTIVE CONTROL PROCESSES:
A GUIDED TOUR

INTRODUCTION

The last years have seen an extraordinary acceleration of interest in the analysis and control of systems with a particular focussing of attention upon self-regulating systems. We must face the fact that we have landed with both feet in the era of automation.

Whether we prefer this newly coined term, or the older—but only slightly older—expression "servomechanism," or think in terms of "feedback control," or employ the neologism of Wiener, "cybernetics," the mathematical transliteration of the problems arising in these studies is the same. A broad spectrum of processes encountered by engineers of all types— electronic, mechanical, industrial, and nuclear—by systems analysts and operations researchers in various phases of industrial, scientific, and military activity, by economists and management consultants, may be subsumed under the following abstract formulation.

Consider a physical system, where the adjective "physical" is used in a sense sufficiently broad to cover any process occurring in the real world. Let us idealize this system sufficiently to be able to describe it at any time t by means of a finite set of quantities $x_1(t), x_2(t), \ldots, x_n(t)$. These quantities, often called "state variables," constitute the components of a vector $x(t)$, the state vector.

In many cases, in an initial study of a control process, it suffices to take $x(t)$ to be one-dimensional, a scalar quantity. Thus, for example, $x(t)$ may represent the position of a particle along a line, the altitude of a satellite, the deviation from equilibrium position of a pointer, or again, the quantity of items in a stockpile, the ratio of an isotope, or the temperature of an oven.

Generally, $x(t)$ will be multidimensional, as in the study of mechanical systems where the components of x are points in phase space (positions and velocities); as in the study of electrical circuits where the components may represent currents and voltages; as in an economic system where the components may be productive capacities and stockpiles of interdependent industries; and finally, as in a biological system where the components may represent the concentrations of a drug in various organs.

In the more complicated study of adaptive control processes, as we shall discuss below, some of the components may be probability distributions.

As we have indicated by our formulation, by the notation and by the examples, we are primarily concerned with dynamic, rather than static,

3

systems. Our interest is in systems that exhibit change over time as a result of the operation of internal and external influences.

The formulation of these processes in analytic terms is far from routine and merits and requires the utmost care and attention. As we shall see in some cases, the very description of a control process in precise terms often requires a certain amount of ingenuity and low cunning—the typical attributes of the mathematician.

It is essential to realize that there are many alternative mathematical approaches to the same problem. One of the prime objects of this book is to indicate a few of the methods that have been successfully employed. Although we do not wish to enter into any detailed discussion of scientific philosophy and mathematical model-building—topics of much fascination in themselves and, oddly enough, of extreme practical significance—we do want to make the following brief remarks. Further discussion at various levels will be found in the various chapters.

A first fundamental question that arises in the construction of descriptive models of the behavior of a physical system is how to relate the time changes in the system to the state of the system. To begin with, assume that we are dealing with a physical process in which time is measured continuously rather than discretely. The change in the system is then measured by the derivative of the state vector dx/dt. A quite simple and extremely useful assumption to make is that this rate of change depends only on the current state of the system and not at all on the past history of the system. This is, of course, always an assumption and never a fact.

This basic assumption leads to a mathematical description of the process by means of a system of ordinary differential equations of the form

$$(i.1) \qquad \frac{dx_i}{dt} = g_i(x_1, x_2, \ldots, x_N, t), \quad x_i(0) = c_i, \quad i = 1, 2, \ldots, N,$$

where the quantities c_i, $i = 1, 2, \ldots, N$ specify the initial state of the system. It is occasionally convenient to employ vector notation and to write, instead of (i.1), the single equation

$$(i.2) \qquad \frac{dx}{dt} = g(x, t), \quad x(0) = c.$$

In many cases hereditary influences play an important role. This means that the assumption that the rate of change depends only upon the current state is not valid. In certain cases we can suppose that the rate of change depends on the current state and the states at certain fixed time-intervals in the past. We then encounter differential-difference equations of the form

$$(i.3) \qquad \frac{dx}{dt} = g(x(t), x(t - t_1), \ldots, x(t - t_k), t).$$

4

These "time-lag" equations arise in many investigations of electrical, mechanical, biological, and economic phenomena in which retardation effects are present. The very use of feedback control introduces these effects.

Most often we are forced to propose even more complicated formulations involving the entire past history of the system. Here the equations may take the form of differential-integral equations

$$(\text{i.4}) \qquad \frac{dx}{dt} = g\Big(x(t), \int_{-\infty}^{t} x(s)\, dG_1(s, t),\, \ldots,\, \int_{-\infty}^{t} x(s)\, dG_k(s, t),\, t\Big),$$

or of integral equations

$$(\text{i.5}) \qquad x(t) = g\Big(\int_{-\infty}^{t} x(s)\, dG_1(s, t),\, \ldots,\, \int_{-\infty}^{t} x(s)\, dG_k(s, t)\Big).$$

Equations of this general type occur in the theory of branching processes and in various parts of the theory of invariant imbedding, in which one considers cosmic ray cascades, biological growth processes, wave propagation, radiative transfer, and so forth. If the equations are linear, the Laplace transform plays a fundamental role in furnishing analytic solutions.

In the following pages we shall consider only systems which can be described either by means of the differential equations of (i.1), or by means of the discrete analogue, the difference equations

$$(\text{i.6}) \qquad x(t + \Delta) = g(x(t), t), \quad x(0) = c,$$

with $t = 0, \Delta, 2\Delta, \ldots$. As we shall see, the nature of modern digital computers makes a description in terms of (i.6) generally vastly preferable to that given by (i.1). Not only are these recurrence relations better suited to the needs of numerical solution, but they also reside on a simpler conceptual and mathematical level. This will particularly be the case when we come to the study of stochastic and adaptive processes. Consequently, they are ideally suited to an expository text of this nature, where it is important to separate the conceptual difficulties inherent in new problems and methods from the well known analytic obstacles.

In the descriptive part of the study of a physical system, a primary problem is that of determining the behavior of the solutions of (i.1) or (i.6) as t increases. This has been a central theme of modern and classical analysis and shows no sign of diminishing in intensity. It is part of the great game of scientific prediction, in which we must prove our understanding of the basic structure of a physical system by predicting its future behavior on the basis of our knowledge of the present state, and perhaps some of the past history.

One of the most important studies of this type is the investigation of the stability of a system. We wish to determine, in advance, whether certain disturbances will jar the system from some preferred equilibrium state, or whether the system will return to this state.

This brings us to the principal objective of these lectures: an investigation of the mathematical aspects of control. In many situations, experience or mathematical analysis will disclose the fact that the operation of the system in its unfettered form is not wholly satisfactory.

It becomes necessary, then, to alter the behavior of the system in some fashion. Before indicating how this may be done, let us give some common examples of phenomena of this nature. A mechanical system such as an ocean liner or a transcontinental airliner may possess dangerous and discomfiting vibrations. These vibrations may cause nausea—a nuisance economically, but a menace militarily—or they may produce elastic fatigue (a hereditary effect, incidentally) that will eventually destroy the ship or plane. An electrical system may dissipate energy in the form of heat. This may be economically wasteful, as in the case of the electric light, or it may be dangerous, resulting in the melting of critical communication cables, with attendant hazard of fire.

An economic system may be consuming scarce or valuable resources without concomitant useful production, or resources potentially productive may be left idle. On one hand there are the threats of depletion and inflation; on the other there is the menace of depression. The control problems in this area are intensified by many retardation effects, and by psychological as well as physical factors, with resultant political overtones.

Finally, consider a biological system such as a human being, where, due to some chemical unbalance, the necessities for a healthy equilibrium may not be produced and drug supplements are prescribed. The administration of too small a dosage leads to illness due to deficiency; too large a dosage will produce harmful side effects. We shall discuss some aspects of medical diagnosis below in connection with adaptive control.

It is abundantly clear from these few examples that control processes play vital roles in all aspects of our existence. The construction of an adequate mathematical theory of control processes thus represents a resounding challenge to the mathematician. As always in the study of fundamental scientific problems, he will be amply rewarded by the many novel, interesting, and significant mathematical problems which he encounters. In return for letting him play his game on his own terms, society will be rewarded by new scientific techniques, and new esthetic and cultural experiences.

Only this constant interchange between the real and abstract—a feedback control process in itself—keeps mathematics vital. Without it there are the dangers of sterility, atrophy, and ultimately, decadence.

Let us now see how to formulate control processes in analytic terms. In order to improve the operation of a system, we can alter its design, bring external influences to bear, or do both. There is no clear line of demarcation between these concepts.

One way or the other, we suppose that the governing equation is no longer (i.1), but an equation of the form

$$(i.7) \qquad \frac{dx}{dt} = g(x, y), \quad x(0) = c.$$

Here $x(t)$ is, as before, the state vector, but $y(t)$ is a new vector, the control vector.

We wish to choose $y(t)$, subject perhaps to restrictions of a type we shall mention below (restrictions which exist in any realistic model of a physical process), in order to have $x(t)$ behave in some prescribed fashion. Two types of processes are of particular importance. In one we wish $x(t)$ to be as close as possible to some prescribed vector function $z(t)$ throughout the duration of the process. Let $h(x(t) - z(t))$ denote some measure of the instantaneous deviation of $x(t)$ from $z(t)$, a scalar function of the components of $x - z$. Then we wish to minimize the functional

$$(i.8) \qquad J(y) = \int_0^T h(x - z) \, dt,$$

a measure of the total deviation of x from the desired state z.

Alternatively, we may not care at all what happens to $x(t)$ for $0 < t < T$, provided only that $x(T)$, the value of x at the end of the process, be a prescribed vector. This is "terminal control."

Problems of this type fall within the province of the calculus of variations, and occasionally these classical techniques can be employed to determine optimal control policies. Specifically, if (i.7) is a linear equation of the form

$$(i.9) \qquad \frac{dx}{dt} = Ax + y, \quad x(0) = c,$$

and $h(x - z)$ is a sum of inner products $(x - z, x - z) + \lambda(y, y)$, a quadratic function of x and y, then explicit analytic solutions can be obtained. These results are often quite useful in furnishing approximations of greater or lesser validity to the solution of more realistic and more significant problems, and thus always represent a first line of attack.

Generally, however, for reasons we shall discuss in some detail below in Chapter 1, the calculus of variations yields little help. We must then develop new techniques in order to obtain solutions to the mathematical problems posed by control processes. Fortunately, to help us in our efforts, we possess a new tool, the digital computer, whose existence changes the very concept of the term "solution."

The new mathematical technique that we shall employ throughout this volume is the theory of dynamic programming. The roots of this theory in

7

classical analysis, and the way in which it can be applied to yield the computational solution of many control problems which defy other techniques, will be part of our story.

Unfortunately, we cannot at this point rest on our laurels. In the first place, within this theory of deterministic control processes there still remain many formidable difficulties to conquer, as we shall see, all centering about the "curse of dimensionality." In the second place, the assumptions that lead to a deterministic equation of the form of (i.7) are, alas, invalid in many significant situations, as we shall discuss in detail. Certainty must yield to uncertainty in our desire for increased accuracy—a paradox!

This last "alas" is to some extent a crocodile tear. For, although the harassed engineers and economists would certainly like to see a cookbook of mathematical recipes for all purposes—a sort of glorified table of integrals, or even of logarithms—the mathematician, in his secret heart, would like to see no such thing. This last named individual is forever delighted to see the close of one investigation afford the entry to a dozen others, the scaling of one height leading merely to the discovery of further mountain ranges. Fortunately, the physical universe very kindly provides him with a hierarchy of problems leading to new theories, and these new theories, to further problems.

The aspects of uncertainty that realistic processes introduce are, at first, elegantly disposed of by the classical theory of probability and the ingenious concept of a stochastic quantity. Upon a suitable mathematical edifice we construct a theory of stochastic control processes which embraces the previous deterministic theory as a special case. In place of (i.7) we envisage the physical system to be governed by a stochastic differential equation of the form

$$\text{(i.10)} \qquad \frac{dx}{dt} = g(x, y, r(t)), \quad x(0) = c,$$

where $x(t)$ is, as before, the state vector, $y(t)$ is the control vector, and $r(t)$ is the stochastic vector representing certain unknown effects.

In order to avoid certain inappropriate analytic complexities inherent in the study of continuous stochastic processes, we consider only discrete stochastic processes, where the differential equation of (i.10) is replaced by the difference equation

$$\text{(i.11)} \qquad x(t + \Delta) = g(x(t), y(t), r(t)), \quad x(0) = c,$$

$t = 0, \Delta, \ldots$.

Here we have no difficulty in defining variational problems of stochastic nature.

We show that these processes may be treated by means of the theory of dynamic programming, using precisely the same techniques that are used in

8

the preliminary chapters. We thus obtain a unified approach to the formulation of deterministic and stochastic control processes and to their computational solutions.

This is, however, not at all the end of the story, but rather the beginning of the most interesting part—a part that will never end. In introducing the random vector $r(n)$, we supposed that its probability distribution was known. In many situations this is not a valid assumption.

One way to overcome the difficulty posed by our ignorance is to introduce the concept of a "game against nature." The unknown distribution for $r(n)$ can be determined on the basis of the strong assumption that it will be chosen in a fashion which minimizes the control we can exert. This idea leads to the mathematical theory of games of Borel and von Neumann. Interesting and stimulating as this theory is, it suffers from a number of defects which force us to look for more satisfactory ways of dealing with uncertainty.

We consider, then, processes in which we can learn about the nature of the unknown elements as the process proceeds. The multistage aspects of the process must be made to compensate for the initial lack of information. We call the mathematical theory developed to treat control processes of this type the theory of adaptive control processes.

Let us consider two particular processes of this nature in order to give the reader some idea of the types of problems that we face. Suppose that we are given two "wonder drugs" for the treatment of a rare disease. The usual approach might be to treat the first hundred or so patients that come into a clinic with the first drug, and the second hundred with the second drug. This procedure, however, may be quite wasteful in terms of human suffering if it turns out that the second drug is very much better than the first. Let us, then, try the first ten on one drug and the second ten on the second drug. On the basis of what happens we may try the next fifteen on the more effective drug and the five patients after that on the less effective drug. Continuing in this way, we can hope to zero-in on the more effective drug more rapidly.

This idea, containing the germ of sequential analysis (developed during the war years 1942–1946 by Wald), was put forth in a little-known paper by W. R. Thompson in 1935.

The problem of determining the optimal testing policy is a problem within the domain of adaptive control processes. Not only in these areas must we determine optimal policies, but we must also determine what we mean by optimal policies. Continually in these new domains we face classes of mathematical problems in which an essential part of the solution is the task of formulation in precise analytic terms.

As we shall see, the theory of dynamic programming yields a systematic technique for formulating problems of this nature and for obtaining computational solutions in certain cases. It is not, however, the only method, and we shall briefly indicate below some of the other new techniques which have been developed. We shall not have time to discuss these other approaches, many of which appear quite promising.

As a second example of an adaptive control process, consider the problem of using a noisy communication channel to determine the original signal emitted by a source. If the statistics of the channel are known, this is a formidable problem—but one within the framework of modern communication theory. If, however, the statistics are unknown initially, we face an adaptive process. Problems of this type can occasionally be treated by application of what is called "information theory."

In the multistage decision processes, which we consider in the parts of the book devoted to deterministic and stochastic processes, we shall assume that we know the state of the system at any time, the set of decisions allowable at any stage, the effects of these decisions or transformations, the duration of the process, and the criterion by which we evaluate the policy that is employed—or distribution functions for all of these quantities. The theory of adaptive control processes is devoted to the study of processes in which some or all of these apparently essential pieces of information are unknown, and unknown in several different ways. We see then that we are impinging on one of the basic problems of experimental and theoretical research—what should we look for, how should we look for it, and how do we know when we have found it?

Finally, let us point out that the theory we present can be used as a basis for the study of automata, robots, and "thinking machines" in general. It is important to note, however, that in all of this the machines do *no* thinking. *We* do the thinking—the machines carry out our policies.

In our concentration upon control processes and related mathematical questions we have avoided contact with the statistical theories of sequential analysis and decision processes in the sense of Wald. The reader interested in pursuing these paths is referred to the books by Wald, as well as to the more recent work of Blackwell and Girshick, Wolfowitz, and others. Closely related to the concept of adaptive control is the stochastic approximation method of Monro and Robbins, another topic we have passed over. References to their work and extensions will be found in the bibliography at the ends of Chapters XIV, XV and XVI.

Another approach to adaptive control processes is through the theory of evolutionary processes due to Box and his colleagues, to which references are also given at the end of Chapter XVI. Neither this nor the deterministic or stochastic versions of the method of steepest descent are discussed.

In our study of multidimensional systems we have passed in silence over the new and intriguing method of Kron, the "tearing method."

To include all of these ideas, stimulating and pertinent as they are, would be to introduce too many skew themes. We have preferred to exploit a particular technique, the theory of dynamic programming, and to show how many different topics can be treated by its means.

Let us now devote a few lines to the specification of the audience for which this book is primarily intended. We hope that it will consist of mathematicians and modern engineers, which is to say, mathematically trained and oriented engineers. To the mathematician we would like to present a vista of vast new regions requiring exploration, where significant problems can be simply stated but require new ideas for their solution. In almost every direction that we have taken, practically everything remains to be done. There are concepts to make precise, arguments to make rigorous, equations to solve, and computational algorithms to derive.

We feel that the theory of control processes has emerged as a mature part of the mathematical discipline. Having spent its fledgling years in the shade of the engineering world, it is now like potential theory, the theory of the heat equation, the theory of statistics, and many similar studies—a mathematical theory which can exist independent of its applications.

To the engineer we offer a new method of thinking about control problems, a new and systematic way of formulating them in analytic terms, and lastly, with the aid of a digital computer, a new and occasionally improved way of generating numerical answers to numerical questions.

Many of the results given here in sketchy outline can be applied to various important types of engineering processes. We have merely indicated a few of the surface applications—those immediately available to a mathematician with a theory who is in search of problems.

We feel that some of the material contained herein should be of interest to the mathematical physicist, particularly the chapters on the dynamic programming approach to the calculus of variations and the new computational techniques thus afforded, and the chapter on quasilinearization. We hope that the chapters on learning and adaptive control processes will be of interest to the theoretical and experimental psychologist, and perhaps to some biologists. Finally, the parts of the book devoted to stochastic and adaptive control processes will be of interest to operations analysts, and mathematical economists.

I should like to thank a number of friends and scientific colleagues for their help in writing this book, both before, during, and after. Particularly, I would like to acknowledge the help of Stuart Dreyfus, (with whom I am otherwise engaged in writing a book on the computational aspects of dynamic programming), for teaching me what I know about the abilities and idiosyncrasies of computers; and the help of Robert Kalaba, with whom I have collaborated in a series of papers on adaptive control processes and invariant imbedding. Without their live enthusiasm and dedication, most of the contents of this book would have remained in the limbo of undreamt dreams.

For selection of some of the snatches of poetry and prose at the beginnings of the chapters and for help in the detection of others, I wish to thank my colleague Betty-Jo Bellman. Her devoted listening and animated commentary have materially aided me in the writing of this book.

I should like once again to acknowledge my appreciation of the untiring aid of my secretary, Jeanette Hiebert. Her patient and rapid typing of draft after draft of the manuscript almost made this writing and rewriting painless.

Finally, I want to express my sincere appreciation of John D. Williams, head of the Mathematics Division at The RAND Corporation. It has been his belief in the value of long-term research which has furnished me the opportunity to work in these new and fascinating areas of science.

CHAPTER I

FEEDBACK CONTROL AND THE CALCULUS OF VARIATIONS

> Darwin and Mendel laid on man the chains
> That bind him to the past. Ancestral gains
> So pleasant for a spell, of late bizarre
> Suggest that where he was is where we are.
>
> DAVID McCORD: *"Progress"*

1.1 Introduction

In this chapter we will discuss a gallimaufry of variational problems of mathematical nature which arise in the course of attempting to treat a variety of control processes of physical origin.

We hope, as far as possible, to lay bare the many approximations that are consciously or unconsciously made in studies of this type, either by force of habit or by dint of necessity. Furthermore, we wish to examine quite closely the many analytic difficulties that attend our path, and to understand what is required of a mathematical solution.

Only if we are very clearly—almost painfully—aware of the manifold aspects of the problems that arise, can we hope to select pertinent mathematical models and utilize meaningful mathematical techniques. As we shall repeatedly emphasize in what follows, concepts play a role equally important with that of equations, and the construction and interpretation of mathematical models is of even greater significance than the solution of the particular equations to which they give rise.

After showing briefly how feedback control processes lead naturally to problems in the domain of the classical calculus of variations, we shall examine in some detail the many obstacles which intervene between analytic formulation and numerical answers This catalogue of catastrophe serves the useful purpose of furnishing a motivation for a new approach to the calculus of variations by way of the theory of dynamic programming, the exposition of which constitutes the substance of two following chapters.

1.2 Mathematical Description of a Physical System

Let us begin with a description of the mathematical model of a physical system which we shall employ here and in the following chapters. We suppose that the physical system under observation varies over time in such a way that all the information that we ever wish to have concerning the system is contained in a set of N functions of time, $x_1(t)$, $x_2(t)$, $\ldots$, $x_N(t)$. When put deliberately in this bald fashion, it becomes clear how crude an approximation this supposition is.

The quantities $x_i(t)$, $i = 1, 2, \ldots, N$ will be called the *state variables*, and N will be called the *dimension* of the system. It is occasionally useful to regard the x_i as components of a vector x, the *state vector*. We write it as a column vector,

$$(1.1) \qquad x = \begin{pmatrix} x_1 \\ x_2 \\ \cdot \\ \cdot \\ \cdot \\ x_N \end{pmatrix}.$$

We now wish to study the behavior of the state vector as a function of time. More ambitiously, given the present state and the past history of the system, we would like to predict the future behavior. Translated into mathematical terms, this means that we would like to determine a function, or more precisely a functional $F(x; \Delta)$, dependent upon $x(t)$ and values of $x(s)$ for $s \leq t$, which will yield the value of $x(t + \Delta)$ for $\Delta > 0$.

Of the many types of functionals available, we shall initially choose those derived from differential equations, for reasons of convenience and tradition. Subsequently, for reasons we shall describe below, we shall employ difference equations, and several sections beyond this we shall indicate how differential-difference equations and partial differential equations enter in quite natural ways.

Having fastened upon differential equations, let us ask, initially, not for the value of $x(t_1)$ at some later time $t_1 > t$, but rather for the instantaneous change in $x(t)$ as a function of its present and past states. We thus write an equation of the form

$$(1.2) \qquad \frac{dx}{dt} = G(x),$$

where $G(x)$ is a functional of the present and past states.

It is clear that both descriptions of a process yield the same information. Either we tell what the future behavior is at any time $t_1 > t$, or we prescribe what the change in state is at any time t. As we shall see, in many situations one formulation may possess many advantages over the other.

In order to simplify the problem by writing a more manageable equation, we boldly assume that $G(x)$ is independent of the past history of the process and depends only upon the present! In other words, we use as our defining equation a *differential equation* of the form

(1.3)
$$\frac{dx}{dt} = h(x),$$

where $h(x)$ is a vector function of x. This equation is shorthand notation for N simultaneous equations of the form

(1.4)
$$\frac{dx_i}{dt} = h_i(x_1, x_2, \ldots, x_N), \quad i = 1, 2, \ldots, N.$$

To appreciate the magnitude of this assumption that we have made, observe that according to well-known mathematical results concerning the existence and uniqueness of solutions of differential equations, under reasonable assumptions concerning the functions $h_i(x_1, x_2, \ldots, x_N)$ we can assert that a system governed by the equations of (1.4) is uniquely specified at any future time t_1 by its state at some previous time $t_0 < t_1$.

In many physical situations this is a manifest absurdity. It is, however, a useful if patent absurdity, and consequently one that we shall base our studies on—for a while. The fault in so many mathematical studies of this type lies not so much in sinning as in a lack of realization of the fact that one is sinning, or even in a lack of acknowledgement of any conceivable type of sin.

A mathematical model of a physical process is to be used as long as it yields predictions in reasonable accordance with observation. Considering the complexity of actual physical phenomena vis-a-vis the simplicity of our mathematical models, it is not surprising that we are forced to modify our formulations from time to time in order to obtain more accurate results. What is remarkable is that deep understanding of many physical processes can be obtained from rudimentary assumptions.

1.3 Parenthetical

Along the lines of what has preceded, we feel that it is worth pointing out that the basic problems of mathematical physics are often not recognized and quite often misinterpreted. Generally speaking, the student gathers the impression from his initial training that the fundamental problem in the theory of differential equations is that of deducing the values of $x(t)$ for $t \geq t_0$, given the initial state of the system and the dynamical equations of (1.4).

This is, of course, a vital problem upon whose solution much depends, but it is not the one that Nature most frequently thrusts at us. Let us discuss this point. Upon examining a physical system and performing various kinds of qualitative and quantitative experiments, we observe various sequences of values.

15

Confronted with this profusion of numbers, the question that faces us is first that of selecting a finite set of state variables, constituting a vector x. Secondly, we must choose a set of functions, constituting the vector $h(x)$, the right-hand side of the differential equation in (1.3). The combination must yield solutions which fit the observed facts, qualitative or quantitative, to within what we may call "experimental error," or, with an accuracy which is sufficient for our purposes.

In other words, the inverse problem is the fundamental one: "Given a set of data, find the equations which will yield these values." It follows that there will be many theories which can be used to explain the same data, particularly if the quantity of data is not too large.

We have posed this problem in terms of ordinary differential equations. In general, the basic equations of mathematical physics are partial differential equations and equations of more complex nature. In the present state of the theory of control processes, it is sufficient to restrict our attention to ordinary differential equations.

1.4 Hereditary Influences

As an example of the inadequacy of mathematical models, we may find that in treating certain processes the assumption that the change in the state is a function only of the current state is an unworkable hypothesis. It may be necessary in order to explain the occurrence of certain phenomena to consider the state of the system at certain times in the past as well as in its present state. In this way we meet *differential-difference* equations,

$$(1.5) \qquad \frac{dx}{dt} = g(x(t), x(t - t_1), \ldots, x(t - t_r)),$$

and the functional equations of more complex form, such as

$$(1.6) \qquad \frac{dx}{dt} = g\left(x(t), \int_{-\infty}^{t} k_1(t, s)x(s)\, ds, \ldots, \int_{-\infty}^{t} k_r(t, s)x(s)\, ds\right).$$

As mentioned above, finite dimensional vectors, in general, do not yield a sufficiently accurate description of physical processes. For example, in treating problems involving heat conduction, we encounter partial differential equations of the form

$$(1.7) \qquad \frac{\partial u}{\partial t} = \frac{\partial^2 u}{\partial x^2} + g\left(u, \frac{\partial u}{\partial x}, x, t\right).$$

Many of the methods we discuss in the following pages can be employed to treat these more complex situations. We shall, however, resist the temptation to explore these new areas.

Let us remark that in some cases theories are forced upon us inexorably by physical "facts." More often, however, the choice of a mathematical formulation is a matter of mathematical and even psychological esthetics.

1.5 Criteria of Performance

The differential equation

(1.8) $$\frac{dx}{dt} = h(x), \quad x(0) = c,$$

as pointed out above, determines the future behavior of the system under consideration, given the initial state c.

It may describe the behavior of a satellite or a rocket, a multivibrator or a cyclotron. Or, in the economic world it may describe the growth of the steel industry, or the fluctuations of our economy. The beauty of mathematics lies in the fact that it furnishes a universal language which characterizes in a uniform way processes of quite different physical appearance. To some extent it transcends the narrow specialization which regrettably and perhaps irresistibly mars our culture, and, to a greater extent, it, together with music, furnishes a bond between people over whose cradles lullabies in many different tongues were sung.

Having passed this hurdle of furnishing a mathematical (i.e. quantitative) model of a physical system, we now wish to evaluate the behavior of the system. Referring to the satellite, we may wish it to pursue a certain periodic orbit. In the case of the multivibrator we want it to maintain a certain fundamental frequency of oscillation, while in the case of the steel industry we may wish to ensure a specified steady growth. On the other hand, as far as the economy is concerned, we may wish to steer a cautious course between inflation and depression.

In some cases, then, we can evaluate performance by comparing the actual state vector $x(t)$ with some preferred state vector $z(t)$. In other cases our evaluation is absolute in terms of the system performance itself.

Quite often we measure the rate of deviation and average this quantity over time. Thus, our criterion of performance may be given by an integral of the form

(1.9) $$\int_0^T Q(x - z) \, dt,$$

where Q is a scalar function of the vector $x - z$.

The most frequently used criterion of this type, for reasons we shall discuss below, is based upon mean-square deviation. Using inner product notation,

(1.10) $$(x - z, x - z) = \sum_{i=1}^N (x_i - z_i)^2,$$

we utilize the time average

(1.11) $$\int_0^T (x - z, x - z) \, dt$$

as a measure of the performance of the system.

17

In some cases this criterion is physically meaningful, furnishing a measure of power loss or dissipation of energy. Generally its appearance is a matter of mathematical convenience and is often necessary if we wish to employ analytic tools and obtain explicit solutions for use as stepping stones to the solution of more complex problems.

On the other hand, for the case of the growth of the steel industry we might agree to measure this by means of an integral of the form

$$(1.12) \qquad \int_0^T K(x)\, dt,$$

a quantity bearing some relation to the actual quantity of steel produced or the total profit realized.

Functionals of $x(t)$ which evaluate the performance of the system will be called *criterion functions*. Sometimes they can be expressed simply in terms of integrals. At other times they are more complicated in structure, and in some quite significant processes they are implicit in nature, as discussed in the following section.

Choice of a criterion function usually involves a compromise between a more accurate evaluation of the physical process and a more tractable mathematical problem.

1.6 Terminal Control

In a number of important instances we can suppose that we have no interest at all in how the system behaves in the course of the process. Instead we are concerned only with the final state of the system. In situations of this type we say that we are interested in *terminal control*.

In many of these cases we can evaluate the process in terms of the state of the system at some *fixed* time T. In this case it is desired to minimize a prescribed function of the terminal state $\phi(x(T))$. In other cases the situation is more complicated, because the time at which we wish to measure the state of the system may not be fixed in advance but may depend in a very critical way upon how the process is carried out. As an example of this, when aiming a space ship at Venus we can evaluate the accuracy of the space trajectory in terms of the minimum miss distance. Here, the time at which the space ship is closest to Venus depends upon the aiming policy. We thus have an example of an *implicit* rather than explicit criterion function. We shall discuss problems of this nature in more detail below. Generally they are rather formidable when approached by classical techniques. We shall instead use dynamic programming methods.

1.7 Control Process

If we find that we do not particularly like the behavior of the system as time proceeds, we can take various steps to alter this behavior. We call this exerting *control*, and a process of this nature is termed a *control process*.

Throughout the remainder of this book we shall study various types of control processes and the many interesting and difficult mathematical processes to which these give rise. Our emphasis will be upon the conceptual, analytic, and computational questions involved, and we shall leave aside all aspects of the actual utilization of the answers that we obtain.

1.8 Feedback Control

Let us now discuss a most ingenious engineering technique which is called *feedback control.*

The basic idea is to use the very deviation of the system from its desired performance as a restoring force to guide the system back to its proper functioning. Schematically,

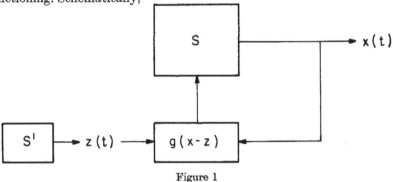

Figure 1

The interpretation of this diagram is the following. S is the actual system with state $x(t)$, and S' is another real or hypothetical system with state $z(t)$. These two states, $x(t)$ and $z(t)$, are compared in one way or another, resulting in a forcing influence $g(x - z)$, which is exerted upon S. This is, of course, the simplest possible block diagram of the application of this technique. Many more complex versions are in current use.

As a result of this "feedback," the equation actually governing the behavior of the system over time has the form

$$(1.13) \qquad \frac{dx}{dt} = h(x, g(x - z)), \quad x(0) = c,$$

replacing the original equation of (1.3). The way the feedback is utilized determines the form of the function g, which, in turn, determines the form of the equation. In many cases the control is applied in such a way that (1.13) takes the simple form

$$(1.14) \qquad \frac{dx}{dt} = h(x) + g(x - z), \quad x(0) = c.$$

19

Applied carefully, the method of feedback control is extremely powerful; applied carelessly, it can actually heighten the effect it was designed to reduce. This is particularly the case when time lags are present, as in processes ruled by the differential-difference equations mentioned above. Here, if feedback control is applied out of phase, even greater deviation and oscillation can result. Examples of this are quite common in the economic world, and many examples are known in mechanical and electrical engineering.

This paradoxical phenomenon is intimately related to the study of the stability of physical systems, a subject which has a venerable and extensive mathematical theory associated with it. On the whole, we will avoid the many questions of stability associated with the computational and analytic algorithms we shall use, in order not to complicate our tale unduly.

Although a certain amount of work has been done in this field, the greater part remains to be done. Most of the difficult and significant problems have barely been touched.

1.9 An Alternate Concept

In place of an approach of this nature, which demands constant surveillance of the system, we can think of an additional extraneous force $y(t)$, determined a priori as a known function of time, applied to the system in such a way as to keep $x(t)$ as close as possible to $z(t)$. Schematically,

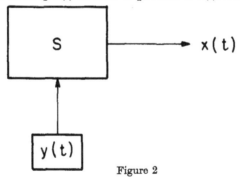

Figure 2

For example, if we are chasing a rabbit which is following an absolutely straight course, we can constantly point towards the rabbit. Here we exert feedback control by using our eyes to guide our motion continuously. Alternatively, having made up our mind that the rabbit is following a straight course, we can refer to our notes from a course in differential equations to determine the curve of pursuit, and then, without further reference to the behavior of the rabbit, we can follow this curve over time.

What is quite surprising is the fact that in the case in which there are no random effects, the *deterministic* case, these two approaches, so different conceptually and physically, are identical mathematically.

This mathematical equivalence enables us to apply a number of powerful analytic tools to the study of variegated feedback control processes. Sometimes one approach is more efficacious, sometimes the other, and sometimes a combination is required. The first approach corresponds closely to what we will subsequently introduce as the dynamic programming

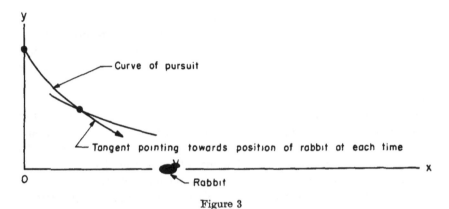

Figure 3

approach, while the second approach is essentially that of the classical calculus of variations. The first approach can readily be extended to treat the stochastic and adaptive control processes discussed subsequently —processes to which the second approach appears completely unsuited.

1.10 Feedback Control as a Variational Problem

Let us now formulate the version of feedback control that we have discussed above as a problem within the framework of the calculus of variations. Having done this, we will sketch the classical formalism and then devote some time to pointing out the formidable difficulties that arise when we attempt to apply these methods to obtain numerical solutions.

Referring to Fig. 1, the problem of exerting optimal control reduces to that of choosing the forcing function $g(x - z)$. Regarding this as an unknown function of t,

$$(1.15) \qquad y(t) = g(x - z),$$

let us write (1.13) in the form

$$(1.16) \qquad \frac{dx}{dt} = h(x, y), \quad x(0) = c.$$

Using the criteria of (1.9) or (1.11), we wish to choose $y(t)$ so as to minimize a functional of the form

$$(1.17) \qquad J(x, y) = \int_0^T k(x, y) \, dt.$$

Here $k(x, y)$ is a prescribed scalar function of x and y.

To treat problems of this nature, we have available the elaborate apparatus of the theory of the calculus of variations. As we shall see, however, despite the powerful techniques that exist for establishing existence and uniqueness theorems, and the varied equations that can be derived, the classical theory provides little help in furnishing numerical solutions unless the functions $h(x, y)$ and $k(x, y)$ have carefully chosen forms. The algorithms that can be used with such effect for demonstrating certain basic properties of the solutions such as existence cannot in the main be utilized for a computational solution.

In view of this, our excursion in the calculus of variations will be brief and purely formal in nature. We shall present the basic variational technique, derive the fundamental Euler equation, and stop short. To justify this cavalier attitude we shall point out in some detail the formidable difficulties strewn along this path to a computational solution.

As we shall see, moreover, in some of the most important applications the classical formalism cannot even be applied, and no Euler equation will exist. These facts, combined with the desire for numerical results, furnish the main impetus to a new approach. The groundwork for this is laid in Chapter II, and the actual discussion is contained in Chapter III. The remainder of the book will, we hope, constitute a justification of this effort.

1.11 The Scalar Variational Problem

To illustrate the fundamental variational approach, it is sufficient to consider the scalar variational problem of minimizing

$$(1.18) \qquad J(u, v) = \int_0^T F(u, v) \, dt$$

over all functions $v(t)$, where u and v are connected by means of the differential equation

$$(1.19) \qquad \frac{du}{dt} = G(u, v), \quad u(0) = c.$$

Here u and v are scalar functions of t.

Proceeding in a purely formal fashion, let $\bar{u}, \bar{v}$ be a pair of functions which yield the minimum, and write

$$(1.20) \qquad \begin{aligned} u &= \bar{u} + \varepsilon w, \\ v &= \bar{v} + \varepsilon z, \end{aligned}$$

22

where ε is a small parameter, and w and z are arbitrary functions of t, defined for $0 \leq t \leq T$.

To obtain the variational equations, we proceed as follows. Write

$$(1.21) \quad J(\bar{u} + \varepsilon w, \bar{v} + \varepsilon z) = \int_0^T F(\bar{u} + \varepsilon w, \bar{v} + \varepsilon z)\, dt$$

$$= J(\bar{u}, \bar{v}) + \varepsilon \int_0^T [w F_{\bar{u}} + z F_{\bar{v}}]\, dt + 0(\varepsilon^2),$$

upon expanding the integrand as a power series in ε. Here, $F_{\bar{u}} = \dfrac{\partial F}{\partial u}$, $F_{\bar{v}} = \dfrac{\partial F}{\partial v}$, evaluated at $u = \bar{u}$, $v = \bar{v}$.

Turning to the differential equation and repeating this process, we have

$$(1.22) \qquad \frac{d\bar{u}}{dt} + \varepsilon \frac{dw}{dt} = G(\bar{u}, \bar{v}) + \varepsilon[w G_{\bar{u}} + z G_{\bar{v}}] + 0(\varepsilon^2).$$

The variational equations obtained by equating the coefficients of ε to zero in (1.21), and to each other in (1.22), are thus

$$(1.23) \quad (a) \qquad \int_0^T [w F_{\bar{u}} + z F_{\bar{v}}]\, dt = 0,$$

$$(b) \qquad \frac{dw}{dt} = w G_{\bar{u}} + z G_{\bar{v}}, \quad w(0) = 0,$$

taken to hold for *all* functions w and z.

The obvious but difficult way to proceed is to solve for w in terms of z, using (1.23b), and then substitute in (1.23a). Much easier is to solve for z in (1.23b), obtaining

$$(1.24) \qquad z = \left(\frac{dw}{dt} - w G_{\bar{u}} \right) \bigg/ G_{\bar{v}},$$

and then to substitute in (1.23a). We thus obtain the relation

$$(1.25) \qquad \int_0^T \left[w F_{\bar{u}} + F_{\bar{v}} \left(\frac{dw}{dt} - w G_{\bar{u}} \right) \bigg/ G_{\bar{v}} \right] dt = 0$$

for all w.

To eliminate the term involving dw/dt, let us use integration by parts:

$$(1.26) \qquad \int_0^T \frac{F_{\bar{v}}}{G_{\bar{v}}} \frac{dw}{dt}\, dt = \left[w \frac{F_{\bar{v}}}{G_{\bar{v}}} \right]_0^T - \int_0^T w \left[\frac{d}{dt} \left(\frac{F_{\bar{v}}}{G_{\bar{v}}} \right) \right] dt.$$

In the expression $[w F_{\bar{v}}/G_{\bar{v}}]_0^T$ the part at $t = 0$ vanishes because of the fixed value of $u(0)$. However, the part at T yields an additional constraint

$$(1.27) \qquad \frac{F_{\bar{v}}}{G_{\bar{v}}} \bigg|_{t=T} = 0.$$

23

Combining (1.25) and (1.26), we obtain the relation

$$(1.28) \qquad \int_0^T \left[\frac{F_{\bar{u}} G_{\bar{v}} - F_{\bar{v}} G_{\bar{u}}}{G_{\bar{v}}} - \frac{d}{dt}\left(\frac{F_{\bar{v}}}{G_{\bar{v}}}\right) \right] w \, dt = 0$$

for all w, which leads to the *Euler equation*

$$(1.29) \qquad \frac{F_{\bar{u}} G_{\bar{v}} - F_{\bar{v}} G_{\bar{u}}}{G_{\bar{v}}} - \frac{d}{dt}\left(\frac{F_{\bar{v}}}{G_{\bar{v}}}\right) = 0.$$

1.12 Discussion

Dropping the bars over the variables u and v, we see that we must solve the system of equations

$$(1.30) \qquad \frac{du}{dt} = G(u, v), \quad u(0) = c,$$

$$\frac{d}{dt}\left(\frac{F_v}{G_v}\right) = \frac{F_u G_v - F_v G_u}{G_v}, \quad [F_v/G_v]_{t=T} = 0.$$

In the special case where

$$(1.31) \qquad \frac{du}{dt} = v, \quad J(u, v) = \int_0^T F(u, u') \, dt,$$

we have the usual equation

$$(1.32) \qquad \frac{d}{dt}(F_{u'}) - F_u = 0,$$

with the usual constraints

$$(1.33) \qquad u(0) = c, \quad F_{u'}|_{t=T} = 0.$$

Observe that one boundary value is at $t = 0$ and the other at $t = T$, a *two-point* boundary-value problem.

There is no difficulty in extending these techniques to treat the multi-dimensional variational problem posed in the preceding section. Since we shall not use these results, there is little point in boring the reader with the details.

1.13 Relative Minimum versus Absolute Minimum

Let us begin our catalogue of difficulties with a caveat which must always be issued when calculus is employed. To observe the phenomenon in its simplest form, consider the problem of determining the minimum value of a function $u = u(t)$ over all real values of t. Suppose that $u(t)$ has the following graph:

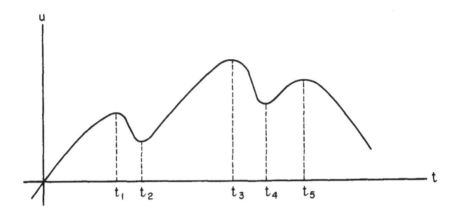

Figure 4

Differentiating and setting the derivative equal to zero, we have the variational equation

(1.34) $$u'(t) = 0,$$

with the solutions

(1.35) $$t = t_1, t_2, t_3, t_4, t_5.$$

Of these solutions three are *relative* maxima, and two are *relative* minima. Of the two relative minima, one is the *absolute* minimum, provided that we take the range of t over a finite interval such as $[a, b]$.

What complicates the matter even further is the existence of *stationary points* of more bizarre nature—points where the derivative is equal to zero, yet which are neither relative maxima nor relative minima. For example, the function $u = t^3$ has neither a relative maximum nor a relative minimum at $t = 0$, although the derivative $3t^2$ is equal to zero there.

As we know, these difficulties compound themselves in the study of the stationary points of functions of several real variables. It is plausible to suspect that when we study functionals or functions of functions, which is to say functions of an infinite number of variables, the problem of determining the nature of a *stationary point* (a point where the first variation vanishes) becomes one of great intricacy.

After all of this is done, assuming that we can distinguish minima and maxima from other types of stationary points, we still face the problem of determining the absolute minimum. As we shall see, the theory of dynamic programming often resolves these difficulties by the simple expedient of

25

bypassing them. There is an old folk-saying to the effect that it is easier to avoid the Devil than to wrestle with him, and the same is true of many mathematical problems.

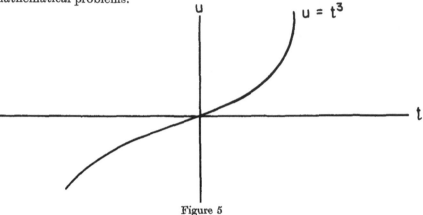

Figure 5

1.14 Nonlinear Differential Equations

Given a system of differential equations, the first thing we do is attempt to find an explicit solution. We know as a result of the expenditure of much effort that this is usually impossible, if by "explicit solution" we mean solution in terms of the elementary functions of analysis and quadratures.

It has, moreover, been known since the work of Liouville that even the second-order linear differential equation

$$(1.36) \qquad u'' + a(t)u = 0$$

cannot be solved in quadratures for all coefficient functions $a(t)$. Furthermore, even a first-order nonlinear differential equation such as the Riccati equation,

$$(1.37) \qquad v' + v^2 + a(t) = 0,$$

simple in appearance as it is, cannot be resolved by quadratures in general. Actually, the two statements are equivalent, since one equation can be transformed into the other by a change of variable of the type

$$(1.38) \qquad u = e^{\int v\, dt}.$$

Since the Euler equations are almost universally nonlinear equations, we must reconcile ourselves to the fact that, apart from a few carefully chosen problems which appear in the textbooks, explicit analytic solutions of variational problems are not to be expected.

If the basic equation is

$$(1.39) \qquad \frac{du}{dt} = au + v, \quad u(0) = c_1,$$

and it is desired to minimize

$$(1.40) \qquad J(u, v) = \int_0^T [k_1 u^2 + 2k_2 uv + k_3 v^2]\, dt,$$

then, as the reader will see upon referring to 1.12, the Euler equation is linear.

Similarly, if the equation governing the system has the vector form

$$(1.41) \qquad \frac{dx}{dt} = Ax + y, \quad x(0) = c,$$

and it is desired to choose y so as to minimize

$$(1.42) \qquad J(x, y) = \int_0^T [(x, B_1 x) + 2(x, B_2 y) + (y, B_3 y)]\, dt,$$

again the variational equation will be linear. These are essential facts when we contemplate the use of successive approximations.

Nevertheless, if we wish to treat significant problems possessing some degree of realism, we must turn our attention to the development of direct computational algorithms, rather than algorithms centering about the Euler equation.

1.15 Two-point Boundary-Value Problems

There is a certain temptation at this point to shrug off these difficulties with the comment that, after all, we do have modern computers at our disposal, and modern computers can certainly treat systems of ordinary differential equations. This is true, but incomplete. Let us analyze the matter in some greater detail.

Differential equations can indeed be solved numerically by means of analogue and digital computers. Although analogue computers are extremely useful when applicable, they can be used only for a very small and specialized class of equations. Let us think then, in all that follows, in terms of digital computers.

To solve any type of functional equation by means of a digital computer, the solution process must be reduced to one involving the basic arithmetic relations of addition, subtraction, multiplication, and division. To treat a scalar differential equation of the form

$$(1.43) \qquad \frac{du}{dt} = g(u, t), \quad u(0) = c,$$

the differential operation must be approximated to by a difference operation. The whole point of this is to reduce a transcendental operation to an arithmetic operation, which can then be performed rapidly and accurately by the digital computer.

The simplest (and perhaps the worst from the standpoint of accuracy) approximation consists of replacing the equation in (1.43) by the difference equation

$$(1.44) \qquad u(t + \Delta) - u(t) = g(u(t), t)\Delta, \quad u(0) = c,$$

where t now assumes only the values $0, \Delta, 2\Delta, \ldots$.

Given the initial value $u(0) = c$, the computer proceeds to determine the values $u(\Delta), u(2\Delta), \ldots$, one after the other, using the recurrence relation in (1.44).

This is not a particularly inspired way of solving an equation of the type appearing in (1.43), but it is feasible because of the tremendous speed of the computer. Having obtained a feasible approach, it is now relevant to ask for more efficient approaches and then for *most* efficient approaches.

At the present time, computational analysis is an art with an associated science, and we do not wish to enter into any of the quite fascinating aspects of this field here. What we do wish to emphasize is that given a differential equation of the foregoing type, there are a number of extremely adroit ways of computing the solution with a high degree of accuracy. Essentially, one can obtain arbitrary accuracy if one is willing to pay the price in computing time.

Consider now a system of two simultaneous differential equations:

$$(1.45) \qquad \frac{du}{dt} = g_1(u, v), \quad u(0) = c_1,$$

$$\frac{dv}{dt} = g_2(u, v), \quad v(0) = c_2.$$

We use exactly the same technique of transforming a set of differential equations into a set of recurrence relations to prepare these equations for a digital computer. Given the initial values, we grind out the subsequent behavior of the solution.

This method carries over to systems of huge dimensions. As a matter of fact, if there is one function that a digital computer can perform and perform well, it is the task of solving systems of nonlinear differential equations with prescribed initial values. As a *routine* application of a new technique, one thinks nothing of exhibiting the solution of forty simultaneous nonlinear differential equations, and even the solution of several hundred simultaneous differential equations is a matter of no great difficulty.

When, however, we turn to the problem of obtaining numerical solutions of nonlinear differential equations subject to two-point boundary conditions, the nature of the situation changes abruptly.

To appreciate the reason for this, consider the second-order scalar equation

$$u'' = g(u, u'),$$

subject to the constraints

(1.47) $$u(0) = c_1, \quad u(T) = c_2.$$

This is called a *two-point boundary-value problem*, since values are prescribed at two distinct points: $t = 0$ and $t = T$.

For a second-order equation, standard computing techniques require a knowledge of u and u' at either the starting point $t = 0$ or the terminal point $t = T$. In the present case we have one value at one point and another at the other.

To use a computer to solve a problem of this nature, we proceed essentially as follows: Guess a value of $u'(0)$, say c_3, and integrate the equation numerically, using the given value of $u(0)$ and the hypothesized value of $u'(0)$. If the value of $u(T)$ obtained in this way agrees sufficiently closely with the required value of $u(T)$, the quantity c_2, we say that we have a solution. If not, we vary the value of c_3 until the agreement is satisfactory.

It is clear that this is not a totally desirable procedure for a number of reasons. In the first place, it is difficult to estimate in advance exactly how long a computation of this type will take. In the second place, it is not clear how accurate it will be. Accuracy at the end-point may not entail equal accuracy throughout the interval $[0, T]$.

Returning to the original variational problem, observe that the equations obtained in 1.11, the Euler equations, are specified by two-point boundary conditions. In addition to the computational difficulties just mentioned, we must face the fact that a computational solution to this equation, following the lines described above, may yield a relative minimum rather than the absolute minimum and, what is an even more distressing thought, may yield a relative maximum rather than a relative minimum.

1.16 An Example of Multiplicity of Solution

As a simple example of how variational problems can possess a set of relative minima, let us examine the problem of determining geodesics on a torus.

Let P and Q be two fixed points on the torus, and consider the problem of determining the curve of minimum length connecting these two points along the surface of the torus. It is clear that there is an *absolute minimum*, indicated on the diagram immediately above by the heavy line joining P and Q.

There is also a minimum distance for all those paths which loop around the ring once before proceeding to Q from P, indicated by the dotted and heavy lines; similarly there is a minimum distance associated with the paths that loop around k times, for $k = 2, 3, \ldots$. These all furnish *relative minima*.

All of these curves will be solutions of the same Euler equation with the same two-point condition. In this case it is easy to see where the relative minima come from. However, in more complicated geodesic problems, like the type encountered in optimal trajectory problems, it is often not

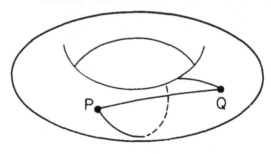

Figure 6

clear in advance whether there are "extraneous" solutions of this type, how many there are, and what they represent physically—if anything at all.

On the other hand, in problems of the Sturm-Liouville type, where we wish to minimize functionals of the form

$$(1.48) \qquad J(u) = \frac{\displaystyle\int_0^T u'^2 \, dt}{\displaystyle\int_0^T g(t)u^2 \, dt} \, ,$$

subject to restrictions of the form $u(0) = u(T) = 0$, the relative minima correspond to the higher modes and play an important role in the original physical process.

1.17 Non-analytic Criteria

In deriving the Euler equation, we have tacitly assumed that the criterion function possesses partial derivatives. Occasionally this may not be the case.

Quite often we wish to determine the control vector y so as to minimize the maximum deviation from equilibrium. Analytically this is expressed by the fact that we wish to minimize the functional

$$(1.49) \qquad \underset{1 \le i \le N}{\text{Max}} \ \underset{0 \le t \le T}{\text{Max}} \ |x_i(t)|.$$

Although this may be the problem whose solution is desired, frequently one sees instead the solution to the problem of minimizing the quadratic functional

$$(1.50) \qquad \int_0^T (x, x)\, dt.$$

This is due to the fact that one problem, the less significant one, can be treated by classical techniques, and the other, the more realistic, cannot.

This recalls the story of the man who is walking down the street at night and sees another man diligently searching at the base of a lamppost. Upon asking whether or not he can help, he is told that a valuable ring was dropped halfway down the street. "Why, then, do you look here under the lamppost?" he asks. "It is lighter here," is the response.

So it is with the solution of variational problems involving linear equations and quadratic criteria. In extenuation let us note that occasionally the solutions to the simpler problems can be used as approximations to the actual solution.

1.18 Terminal Control and Implicit Variational Problems

So far, we have discussed variational problems which, although difficult in many respects, still possess the merit of exhibiting explicit functionals to minimize. In a number of important applications which have arisen in recent years, we encounter *implicit* rather than explicit functionals, as already noted in 1.6 where terminal control was discussed.

An interesting example of this occurs in connection with the problem of landing a rocket ship on the moon. At the time of contact we would like the velocity of the ship to be as small as possible. The terminal velocity, however, depends upon the time of landing, which in turn depends upon the control policy used. Consequently, the upper limit T appearing in the previous functional is itself now a functional.

Another example of an implicit criterion comes under the heading of "bang-bang" control. Consider a system which was designed to be stable and which is indeed stable. A plot of the deviation from equilibrium value (taken to be zero) of one of the state variables may have the following form:

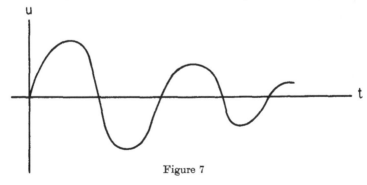

Figure 7

31

For the sake of definiteness suppose that the function is given by the expression

(1.51) $$u = e^{-rt} \cos bt,$$

with $r > 0$.

Consequently, there will be no danger of u increasing arbitrarily in size as t increases, and clearly, the amplitude of the peak deviation over a period actually tends to zero as time increases. However, if r is small the rate of decrease may be too slow for efficient performance. For practical purposes the system may appear to possess periodic fluctuations. This oscillation may result in loss of accuracy, heating, or other waste of energy, or it may cause physical discomfort due to seasickness or airsickness.

Redesigning the system may be impractical, or impossible, or of little avail, because of lack of understanding of the physical phenomena involved. We may observe the effects but not be able to isolate the causes. In a situation of this nature we try to alleviate the condition by use of feedback control exerted in such a way as to minimize the time required to drive the system into the desired equilibrium state. This is another example of a terminal control process without an explicit criterion function.

The name *bang-bang control* is often given to this type of process, since it is frequently desirable to use only the simplest kind of control mechanism—one which operates at a constant level either in one mode or another.

Analytically, the problem may be posed in the following form. Given the vector equation

(1.52) $$\frac{dx}{dt} = g(x, y), \quad x(0) = c,$$

with the equilibrium state $x = 0$, provided that $y = 0$, we wish to determine the control vector $y(t)$, subject to the constraint that

(1.53) $$|y_i(t)| \leq m_i, \quad i = 1, 2, \ldots, N,$$

or, sometimes, that

(1.54) $$y_i(t) = \pm m_i,$$

so as to minimize the time required to drive the solution from its initial state c to the desired state 0.

We shall discuss this problem again in a later chapter. In the meantime let us focus our attention on a new phenomenon, the relations in (1.53) and (1.54).

1.19 Constraints

The difficulties described in the preceding sections remain, and formidable new ones are introduced, if we impose realistic constraints on the types of admissible control policies and admissible states of the system.

We meet a hint of this when we consider the simpler problem of minimizing a function $u = u(t)$ over a closed interval $[a, b]$. As we have already

32

noted, the usual variational technique furnishes the stationary points of $u(t)$. Nonetheless, the actual minimum value of $u(t)$ may be at one of the endpoints of the interval, as in the figure below:

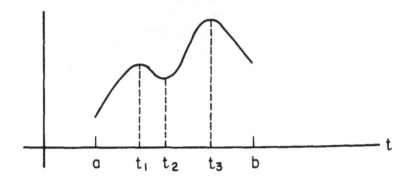

Figure 8

The points t_1, t_2, t_3 are relative extrema; a is the point yielding the absolute minimum.

To illustrate what effect constraints can have in the calculus of variations, consider the relatively simple problem of minimizing the functional

$$(1.55) \qquad J(u) = \int_0^T g(u, u') \, dt,$$

subject to the constraints

$$(1.56) \quad (a) \qquad u(0) = c_1,$$
$$(b) \qquad |u'(t)| \leq c_2, \quad 0 \leq t \leq T.$$

As a function of t, $u'(t)$ may have the form

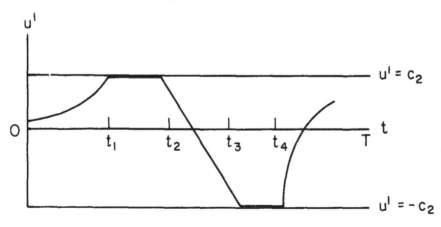

Figure 9

It is easy to construct mathematical examples of this phenomenon of oscillation between extremes of control, and many physical processes can be found where the optimal control policy is of this nature.

In attempting to treat a problem of this type by means of variational techniques, we find that we can only obtain an Euler equation when a free variation is permitted. A free variation is permitted only when, as indicated above, t is between 0 and t_1, t_2 and t_3, and t_4 and T. In the other intervals, $[t_1, t_2]$, $[t_3, t_4]$, only a one-sided variation is permitted. The result is that we obtain Euler inequalities in these intervals. Once again we shall not enter into these matters, since we shall pursue an entirely different route in our discussion of these processes. References to papers containing analytic treatment of questions of this nature will be found at the end of the chapter.

At the present time there exists no uniform method for solving problems of this nature. We know neither how to determine the number of different parts of the solution, the location of the transition points, t_i, nor how to piece these parts together, without imposing special conditions on the form of the differential equation and the form of the criterion function. This is an important and fascinating problem which deserves some attention.

1.20 Linearity

Although nonlinearity is an inconvenience when it comes to the explicit solution of variational problems, lack of it can be an embarrassment too. If all of the equations and functionals that appear are linear, then the classical variational techniques again fail. There are now no Euler equations.

Problems of this type arise most often in the domain of mathematical economics in the study of economic and industrial control processes. Only occasionally in engineering control processes does one encounter an embarrassment of linearity.

Typical of the mathematical questions that one encounters is the following:

"Given the vector-matrix equation

$$(1.57) \qquad \frac{dx}{dt} = Ax + By, \quad x(0) = c,$$

determine the vector y, subject to constraints of the form

$$(1.58) \qquad Cx + Dy \leq f, \quad 0 \leq t \leq T,$$

so as to maximize the functional

$$(1.59) \qquad J(y) = (x(T), a) + \int_0^T (x, b) \, dt."$$

Some problems of this type can be resolved analytically using the Neyman-Pearson lemma in place of the classical techniques; some can be treated by means of the functional equation technique of dynamic programming; and some can be resolved computationally using linear programming techniques, either discrete or continuous versions. Once again, it cannot be said that any uniform method of solution exists for problems of this genre. This is also an area meriting considerable research.

1.21 Summing Up

Our aim in this chapter has been to introduce the concept of feedback control and to show how, under appropriate assumptions, it leads to various classes of variational problems.

Although many of these problems are within the domain of the calculus of variations, a closer examination shows that the classical techniques cannot be readily applied to yield either numerical or analytic solutions.

Our purpose, then, is to introduce a new approach which can be applied to obtain computational solutions of a number of these problems and, in some cases, to obtain the analytic structure of the solution as well. Furthermore, as we shall show in the second and third parts of this volume, this new mathematical framework will enable us to treat stochastic control processes and, finally, adaptive control processes.

In continually emphasizing the need for techniques yielding numerical solutions, our objective has not at all been to play down the role of analytic investigation. Rather, we have tried to stress an idea which is as yet fairly novel, namely the role of the computer in *mathematical experimentation*.

When examining a control process of a new variety which remains stubbornly resistant to the mathematical tools currently at our disposal, it is comforting to know that we may possess systematic means for grinding out the optimal control policies. Carrying this out for various values of the parameters, we hope to detect a pattern in solution which will enable us to discern the structure of the optimal policies. With clues furnished by numerical work, corresponding to those furnished the theoretical physicist by the experimental physicist, we can hope to blaze analytic trails into new regions.

Apart from this, there is another advantage to having exact numerical solutions available. Often, if the precise solution is quite complicated in nature, it is more useful by far to have simple approximate solutions than the actual exact solution. Without comparison with the exact solution, however, it may be difficult to judge the value of a proposed approximation.

Generally, in applications of mathematics to physical processes, there are several distinct phases which we can recognize. The first is that of formulation in mathematical terms—often the most difficult. The second involves an examination of the equations resulting from this formulation to see whether or not they are too imprecise or too rigid. In the first place, there may exist many possible solutions; in the second place, there may exist no solutions. This involves the existence and uniqueness of solutions of the describing functional equations. We enter the third phase when we turn to the problem of obtaining algorithms which will produce the solution, and the fourth phase is that of deriving feasible and efficient algorithms.

Usually the fourth stage will succeed only if we possess a detailed understanding of the structure of the solution. Often this will require a reformulation of the original problem in different mathematical terms.

Only when we comprehend the nature of the solution, from both the mathematical and physical standpoints, and possess feasible and effective algorithms for numerical solution, can we consider a problem solved.

Bibliography and Discussion

1.1 A fascinating introduction to various aspects of the theory of control processes will be found in the books

N. Wiener, *Cybernetics*, Wiley, 1948.

W. S. McCulloch, *Finality and Form*, C. C. Thomas, 1952.

For a discussion of the mathematical aspects of feedback control applied to linear systems, see

J. G. Truxal, *Automatic Feedback Control System Synthesis*, McGraw-Hill, 1955,

D. Middleton, *An Introduction to Statistical Communication Theory*, Wiley, 1960,

where many further references will be found.

One of the first papers on the mathematical theory of control processes (for knowledge of which I am indebted to R. Kalaba) is

J. Clerk Maxwell, "On governors," *Proc. Royal Soc. London*, vol. 16, 1867, pp. 270–283.

This was stimulated by the governor on the Watt steam engine and other analogous devices.

The study of governors, however, far antecedes Watt. According to the beautifully written review article by H. Bateman, "The control of an elastic fluid," *Bull. Amer. Math. Soc.*, vol. 51 (1945), pp. 601–646, a centrifugal governor was invented by C. Huygens for the regulation of windmills and water wheels in the seventeenth century. Many useful references will be found in this paper.

Independently, the mathematical theory of control was initiated by N. Minorsky; see

N. Minorsky, *J. Amer. Soc. Naval Engineers*, May 1922, and May 1930.

These papers discussed the theory and experimental results concerning the use of automatic steering on board the U.S.S. New Mexico.

For discussions of some of the philosophical and scientific problems encountered in setting up mathematical models of physical processes, see

E. Wigner, The unreasonable effectiveness of mathematics in the natural sciences, *Comm. Pure and Applied Math.*, Vol. XIII (1960), pp. 1–14.

J. Von Neumann, "The Mathematician", pp. 180–197 of *Works of the Mind,* Univ. of Chicago Press, 1947.

R. Bellman and P. Brock, "On the concepts of a problem and problem-solving," *Amer. Math. Monthly,* Vol. 67 (1960), pp 119–134.

R. Bellman, C. Clark, C. Craft, D. Malcolm, and F. Ricciardi, "On the construction of a multi-person, multi-stage business game," *Operations Research,* vol. 5, 1957, pp. 469–503.

1.2 The requisite existence and uniqueness theorems for differential equations may be found in

R. Bellman, *An Introduction to the Stability Theory of Differential Equations,* McGraw-Hill, 1953.

S. Lefschetz, *Differential Equations: Geometric Theory,* Interscience, 1957.

Whereas continuity of $g(x)$ in the neighborhood of $x = c$ is sufficient to ensure the existence of at least one solution of $dx/dt = g(x)$, $x(0) = c$, it is not a strong enough condition to guarantee uniqueness. For this, one needs essentially a Lipschitz condition, $\|g(x) - g(y)\| \leq k_1\|x - y\|$ in the neighborhood of c. Slightly weaker conditions will suffice.

1.3 Inverse problems of this type play a far greater role than is generally realized. For the way in which one arises in quantum mechanics, see

M. Verde, "The inversion problem in wave mechanics and dispersion relations," *Nuclear Physics,* vol. 9, 1958–1959, pp. 255–266.

More mathematical presentations will be found in

G. Borg, "Eine umkehrung der Sturm-Liouvilleschen Eigenwertaufgabe. Bestimmung der differential Gleichung durch die Eigenwerte," *Acta Math.,* vol. 78, 1946, pp. 1–96.

I. M. Gelfand and B. M. Levitan, "On the determination of a differential equation from its spectral function," *Translations Amer. Math. Soc.,* vol. 2, 1955, pp. 253–304.

1.4 A survey of the way in which various physical processes can give rise to functional equations of this type may be found in

R. Bellman and J. M. Danskin, *A Survey of the Mathematical Theory of Time-lag, Retarded Control and Hereditary Processes,* The RAND Corporation, Report R-256, April 14, 1954.

An expanded version of this will soon appear in a book written in collaboration with K. L. Cooke. See also

A. D. Myskis, *Linear Differential Equations with Retarded Argument,* Moscow, 1951; German translation: Deutscher Verlag, Berlin, 1955.

R. Aris, "On Denbigh's Optimum Temperature Sequence," *Chem. Eng. Science,* Vol. 12 (1960), pp. 56–64.

For later references, see

N. H. Choksy, "Time-delay Systems—a Bibliography," *IRE Trans. PGAC,* Vol. AC-5, 1960, pp. 66–70.

The first mathematician to make intensive study of hereditary influences in mathematical physics of the kind that occur in elasticity and magnetic hysteresis was V. Volterra. See

V. Volterra, *Leçons sur les Equations Integrals et les Equations Integro-differentielles*, Gauthier-Villars, Paris, 1913.

For a discussion of scientific theories, see

E. P. Wigner, "The unreasonable effectiveness of mathematics in the natural sciences," *Comm. Pure and Applied Math.*, vol. XIII, 1960, pp. 1–14.

Other references will be found there.

1.5 For discussions of the ways in which various economic processes lead to control processes, see

R. Bellman, *Dynamic Programming*, Princeton University Press, 1957.

K. Arrow, S. Karlin, and H. Scarf, *Studies in the Mathematical Theory of Inventory and Production*, Stanford University Press, 1958.

For a detailed discussion of criterion functions, see

C. W. Merriam, III, "An optimization theory for feedback control design *Information and Control*, vol, 3. 1960, pp. 32–59.

1.6 In subsequent pages we will discuss particular terminal control processes arising in the treatment of trajectory problems, in chemical engineering processes, and in servo-mechanism theory in the form of the "bang-bang" control problem.

1.8 The first one to point out the influence of time-lags in control processes and stability theory was N. Minorsky. See

N. Minorsky, Self-excited oscillations in dynamical systems possessing retarded action, *J. Applied Mech.*, vol. 9, (1942), pp. 65–71.

For an introduction to the mathematical theory of stability, see

J. J. Stoker, *Nonlinear Vibrations in Mechanical and Electrical Systems*, Interscience, 1950.

R. Bellman, *Stability Theory of Differential Equations*, McGraw-Hill, 1954.

An important list of papers on adaptive control is contained in

P. R. Stromer, "Adaptive or self-optimizing systems—a bibliography," *IRE Trans. on Automatic Control*, vol. AC-4, 1959, pp. 65–68.

See also

R. C. K. Lee, L. T. Prince, and R. N. Bretoi, "Adaptive control, new concept in automatic flight," *Space Aeronautics*, Feb., 1959, pp. 128–131.

J. A. Aseltine, A. R. Mancini, and C. W. Sarture, *Trans. IRE*-PGAC-101, April, 1958.

For an account of recent Russian work in the theory of control, see

J . P. LaSalle and S. Lefschetz, *Recent Soviet Contributions to Ordinary Differential Equations and Nonlinear Mechanics*, RIAS Technical Report 59-3, AFOSR Doc. No. TN-59-308, 1959.

1.9 As we shall see in a later chapter, this equivalence is a consequence of the dual properties of Euclidean space, in which a curve may be regarded as a locus of points or an envelope of tangents.

For an interesting account of the history of pursuit processes, with many references, see

A. Bernhart, "Curves of pursuit—II," *Scripta Math.*, vol. XXIII, 1957, pp. 49–65.

Further references are given at the end of Chapter 14; see the comments in 14.12.

1.11 Accounts of the classical calculus of variations, which we shall bypass almost entirely, may be found in

R. Courant and D. Hilbert, *Methoden der Mathematischen Physik*, Interscience, 1953.

R. Weinstock, *The Calculus of Variations*, McGraw-Hill, 1956.

O. Bolza, *Variationsrechnung*, Leipzig–Berlin, 1909.

1.12 Other accounts of the many difficulties associated with classical variational techniques are given in

R. Bellman, "Dynamic programming and the numerical solution of variational problems," *Operations Research*, vol. 5, 1957, pp. 277–288.

————, "Dynamic programming and the computational solution of feedback design control processes," *Proc. Computers in Control Systems Conference*, AIEE Special Publ. T-101, Atlantic City, October, 1957, pp. 22–25.

1.14 For a presentation of Liouville's theory in modern, rigorous fashion, see

J. F. Ritt, *Integration in Finite Terms*, Columbia University Press, 1948.

For an application to the integration of the Bessel equation in terms of elementary transcendentals, see

G. N. Watson, *Bessel Functions*, Cambridge Univ. Press, 1944, pp. 111–123.

A new approach to the study of nonlinear functional equations of all types is given in

R. Kalaba, "On nonlinear differential equations, the maximum operation, and monotone convergence," *J. Math. and Mech.*, vol. 8, 1959, pp. 519–574.

1.15 Further discussion may be found in

R. Bellman, "Dynamic programming, invariant imbedding, and two-point boundary value problems," *Boundary Problems in Differential Equations*, Univ. of Wisconsin Press, 1960, pp. 257–272.

1.18 A number of references to recent American and Russian work on control processes of this type will be found at the end of Chapter 4, in the comments on 4.15.

1.19 Analytic approaches via the classical variational techniques to problems involving constraints may be found in

R. Bellman, W. H. Fleming, and D. V. Widder, "Variational problems with constraints," *Annali di Matematica*, ser. 4, tomo 41, 1956, pp. 301–323.

R. Bellman, I. Glicksberg, and O. Gross, "On some variational problems occurring in the theory of dynamic programming," *Rend. del Cir. Mate. di Palermo*, ser. 2, tomo 3, 1954, pp. 1–35.

A. Miele and J. O. Cappellari, *Topics in Dynamic Programming*, Purdue University, July, 1958.

1.20 For the "simplex technique" applied to the problem of maximizing linear forms subject to linear constraints, see

G. Dantzig, The generalized simplex method . . . , *Pac. J. Math.* vol. 5 (1955) pp. 183–195.

For a continuous version of this method applicable to the problem described above, see

S. Lehman, *On the Continuous Simplex Method*, The RAND Corporation, Research Memorandum RM-1386, December 1, 1954.

For analytic solutions of problems of this nature, see

R. Bellman, *Dynamic Programming*, Princeton University Press, 1957.

R. Bellman and S. Lehman, *Studies in Bottleneck Processes* (unpublished).

R. Bellman, I. Glicksberg, and O. Gross, *Some Aspects of the Mathematical Theory of Control Processes*, The RAND Corporation, Report R-313, 1958.

For some problems of this type arising in chemical engineering, see

O. Bilous and N. R. Amundsen, *Chem. Eng. Sci.*, vol. 5, 1956, pp. 81–92, pp. 115–126.

H. P. F. Swinnerton-Dyer, "On an extremal problem," *Proc. London Math. Soc.*, vol. 7, 1957, pp. 568–583.

We have omitted, in what follows, any discussion of the many fascinating and important problems arising in biological and medical studies. For some biological questions, see

V. Volterra, *La Lutte pour la Vie*, Gauthier-Villars, Paris, 1906.

R. Bellman and R. Kalaba, "Some Mathematical Aspects of Optimal Predation in Ecology and Boviculture," *Proc. Nat. Acad. Sci.*, 1960.

A comprehensive book on the various control processes operating in human physiology is

W. B. Cannon, *The Wisdom of the Body*, W. W. Norton, New York, 1932. The author uses the term "homeostasis" to denote the equilibrium states of living organisms. The comments on pp. 24–25 of his book are quite interesting. The substance of these remarks is that conventional feedback control theory may gain from the study of physiological processes, rather than vice versa.

Let us also quote the comment of C. Bernard, on p. 38.

". . . all the vital mechanisms, however varied they may be, have only one object, that of preserving constant the conditions of life in the internal environment."

A number of interesting control processes arise in the administration of drugs in chemotherapy. See

R. Bellman, J. Jacquez, and R. Kalaba, "Some mathematical aspects of chemotherapy—I: one organ models," *Bull. Math. Biophysics*, to appear.

R. Bellman, J. Jacquez, and R. Kalaba, "The distribution of a drug in the body," *Bull. Math. Biophysics*, to appear.

The study of biological and medical control processes is in its barest infancy. These fields represent two of the most challenging and significant scientific arenas of our time. So complex and baffling are the problems that one faces that it is by no means certain that mathematical techniques will succeed. Nevertheless, it is necessary to pursue these investigations with the utmost perseverance.

CHAPTER II

DYNAMICAL SYSTEMS
AND TRANSFORMATIONS

> Who has taught us the true analogies, the
> profound analogies which the eyes do not see,
> but which reason can divine? It is the mathe-
> matical mind, which scorns content and clings
> to pure form.
>
> HENRI POINCARÉ
> *"Analysis and Physics"*

2.1 Introduction

To prepare the way for the application of the functional equation
technique of dynamic programming to the study of control processes, it
is necessary for us to review the classical interpretation of the behavior
over time of a dynamical system as a continuous sequence of transformations.

We particularly wish to emphasize the fundamental consequence of the
uniqueness theorem for the solutions of differential equations, since it is
this which yields the semigroup property which is the law of causality.
All of these concepts generalize to variational processes, as we shall see.

To illustrate the application of some of these ideas in their simpler form,
we shall show how they lead to the functional equations for the exponential
and trigonometric functions. Finally, in order to familiarize the reader
with the functional equation technique, we shall consider some problems of
maximum range and maximum altitude connected with rockets and space
travel, and with trajectory problems in general.

2.2 Functions of Initial Values

Let us consider a system of differential equations of the form

$$(2.1) \qquad \frac{dx}{dt} = g(x), \quad x(0) = c,$$

a vector equation. Here x is an N-dimensional vector function of t, $g(x)$
is a vector function of x, and the vector c represents the initial state of the
system. As before, we assume that $g(x)$ satisfies conditions sufficient to
ensure that there exists a unique solution for $t \geq 0$.

Customarily one writes $x = x(t)$ indicating the dependence of the solution upon t, usually the time. We now make the crucial observation that the solution is also a function of c, the initial value. Hence, we write

(2.2) $$x = x(c, t).$$

This idea was extensively used by Poincaré, and has played a vital role ever since in the study of nonlinear differential equations. What it implies is that the study of particular solutions of differential equations may best be carried out by studying families of solutions.

This concept carried over to the study of other types of functional equations forms the basis of the theory of *dynamic programming*, and, directed toward the field of mathematical physics, constitutes the keystone of the theory of *invariant imbedding*.

2.3 The Principle of Causality

Let us now state a simple consequence of the existence and uniqueness postulated for the solution of (2.1). Suppose that we examine the solution at time $s + t$. Since this solution is *unique*, we obtain the same state if we allow the system to unravel for a time s, observe the state at this time, and then allow it to operate for an additional time t starting with the value attained at time s.

Schematically,

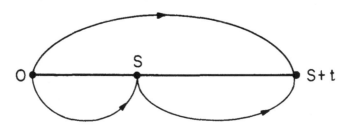

Figure 10

The equality between the two final states arrived at in this way is, of course, the determinism of classical physics, the "principle of causality."

2.4 The Basic Functional Equation

From this verbal statement we readily obtain one of the basic functional equations of mathematics. Expressing the equality of the final states at time $s + t$, derived in these two different fashions, we have the relation

(2.3) $$x(c, s + t) = x[x(c, s), t].$$

This equation expresses the semi-group property of the solutions of the differential equation (2.1).

42

2.5 Continuous Version

Consider the special case in which s is taken to be an infinitesimal. Then, proceeding formally,

$$(2.4) \qquad x(c, s) = c + sg(c) + O(s^2),$$

$$x(c, s + t) = x(c, t) + s\frac{\partial x}{\partial t} + O(s^2).$$

The functional equation (2.3) then yields the relation

$$(2.5) \qquad x(c, t) + s\frac{\partial x}{\partial t} + O(s) = x[c + sg(c) + O(s), t]$$

$$= x(c, t) + s\sum_{i=1}^{N} g_i(c)\frac{\partial x}{\partial c_i} + O(s),$$

or, upon passing to the limit, the linear partial differential equation

$$(2.6) \qquad \frac{\partial x}{\partial t} = \sum_{i=1}^{N} g_i(c)\frac{\partial x}{\partial c_i}.$$

Here the quantities $g_i(c)$ are the N components of $g(c)$.

It is interesting to see how these ideas yield the well-known connection between the original *nonlinear* system of differential equations and a *linear* partial differential equation. Subsequently, we shall obtain a generalization of this relation, valid for variational processes. The generalization, however, will be nonlinear.

2.6 The Functional Equations Satisfied by the Elementary Functions

The relation in (2.3) is the origin of the functional equations satisfied by the elementary functions of analysis. Consider, for example, the scalar function $u(t) = e^{at}c$ defined as the solution of the scalar equation

$$(2.7) \qquad u' = au, \quad u(0) = c.$$

Using (2.3), we have

$$(2.8) \qquad e^{a(s+t)}c = e^{at}(e^{as}c),$$

which yields the functional equation for the exponential function,

$$(2.9) \qquad e^{(s+t)} = e^t e^s.$$

Similarly, consider the function $u = \sin t$, defined as the solution of

$$(2.10) \qquad u'' + u = 0, \quad u(0) = 0, \quad u'(0) = 1.$$

Since $\sin(s + t)$ is a solution for any s, and since $\cos t$ and $\sin t$ constitute a set of fundamental solutions, we must have a relation of the form

$$(2.11) \qquad \sin(s + t) = b_1 \cos t + b_2 \sin t,$$

where b_1 and b_2 are functions of s. Setting $s = 0$, we have

$$(2.12) \qquad b_1 = \sin s.$$

Differentiation, followed by setting $s = 0$, yields

$$(2.13) \qquad b_2 = \cos s.$$

Thus, the addition formula,

$$(2.14) \qquad \sin(s + t) = \sin s \cos t + \cos s \sin t.$$

This, and the result of (2.9), are both special cases of the functional equation for the matrix exponential which we derive below.

2.7 The Matrix Exponential

Finally, consider the matrix equation

$$(2.15) \qquad \frac{dX}{dt} = AX, \quad X(0) = C,$$

whose solution may be written

$$(2.16) \qquad X = e^{At}C.$$

Use of (2.3) yields the equation

$$(2.17) \qquad e^{A(t+s)}C = e^{At}(e^{As}C),$$

or the basic relation

$$(2.18) \qquad e^{A(t+s)} = e^{At}e^{As}.$$

This is a *derivation* of the result, as opposed to the usual proofs which are *verifications*.

Consideration of equations such as (2.15) for the case where A is a linear partial differential operator yields corresponding functional equations for the basic functions of mathematical physics, such as the Bessel functions, Legendre polynomials, and so on. This elegant method for deriving fundamental properties of these functions is due to Hadamard.

2.8 Transformations and Iteration

Returning to (2.1), let us set $t = 1$, and write

$$(2.19) \qquad x(c, 1) = y(c).$$

In other words, we regard the value of $x(t)$ at the end of a unit of time as a transformation $y(c)$ of the initial state c.

Then,

(2.20) $$x(c, 2) = x[x(c, 1), 1] = x[y(c), 1] = y[y(c)],$$

and generally,

(2.21) $$x(c, n) = y^{(n)}(c),$$

where $y^{(n)}(c)$ represents the n-th *iterate* of the function $y(c)$.

It follows that the behavior of the solution $x(c, t)$ as a function of t be regarded as equivalent to the study of the successive iterates of a specific function of c. This idea, again due to Poincaré, was intensively developed by Hadamard, Birkhoff, and others. Using this identity between specific transformations and the behavior over time of various physical systems, due to B. O. Koopman and Carleman, the methods of modern abstract analysis were applied to the study of the famous ergodic hypothesis by von Neumann, Birkhoff, and Carleman.

In what follows, we shall frequently study only the discrete versions of dynamical systems, described by difference equations, or recurrence relations, of the form

(2.22) $$x_{n+1} = g(x_n), \quad x_0 = c.$$

The study of the behavior of the solution as n increases is precisely the study of the iterates of the transformation $y = g(x)$.

This concept of transformations of initial state will play a paramount role in all of our subsequent treatment of control processes. As we shall see, not only does it furnish us a new approach to the calculus of variations, but it is also helpful in treating some descriptive processes, as in 2.10–2.14.

2.9 Carleman's Linearization

While on the subject of differential equations and transformations, it is worth noting an observation of Carleman to the effect that the study of any *nonlinear* system of differential equations can always be shown to be equivalent to the study of an infinite system of *linear* equations.

Consider, for example, the equation

(2.23) $$u' = u^2 + 1, \quad u(0) = c_1,$$

and write

(2.24) $$v_n = u^n.$$

Then

(2.25) $$\frac{dv_n}{dt} = nu^{n-1}\frac{du}{dt} = nu^{n-1}(u^2 + 1)$$
$$= nv_{n+1} + nv_{n-1}, \quad n = 1, 2, \ldots,$$

the desired infinite system of linear equations.

In order to obtain the solution of (2.23), we affix the initial conditions

$$(2.26) \qquad v_n(0) = c_1^n, \quad n = 1, 2, \ldots .$$

Similar techniques can be applied to many other types of nonlinear functional equations. Cutting off the infinite term in (2.25) at some finite value of n, we obtain a finite linear system which yields an approximate solution to the original nonlinear equation.

2.10 Functional Equations and Maximum Range

The current interest in rockets and space travel has aroused a corresponding interest in the determination of maximum range, minimum time, and so on, for various types of trajectories. Here, we wish to show how the concept of dependence of the state vector upon initial state discussed above can be used to obtain computational schemes for determining quantities such as maximum altitude, range, and so on, which seem more efficient than those currently used, and which occasionally can be used to obtain direct analytic results.

2.11 Vertical Motion—I

Consider an object, subject only to the force of gravity and the resistance of the air, which is propelled straight up. In order to illustrate the technique, we shall employ, in simplest fashion, let us treat first the problem of determining the maximum altitude that this object attains.

Let the defining equation be

$$(2.27) \qquad u'' = -g - h(u'),$$

for $u > 0$, with the initial conditions $u(0) = 0$, $u'(0) = v$. Here $v > 0$, and $h(u') \geq 0$ for all u'.

Since the maximum altitude is a function of v, let us introduce the function

$$(2.28) \qquad f(v) = \text{the maximum altitude attained, starting} \\ \text{with initial velocity } v.$$

From the definition of the function, it follows that

$$(2.29) \qquad f(v) = v\Delta + f(v - [g + h(v)]\Delta) + O(\Delta^2),$$

for Δ an infinitesimal.

Verbally, this states that the maximum altitude is the altitude gained over an initial time Δ, plus the maximum altitude attained starting with a velocity $v - [g + h(v)]\Delta$, (the velocity of the object at the end of time Δ) to within terms which are $o(\Delta)$.

Expanding both sides and letting $\Delta \to 0$, the limiting form of (2.29) is the equation

$$(2.30) \qquad f'(v) = \frac{v}{g + h(v)} .$$

Since $f(0) = 0$, this yields the expression

$$(2.31) \qquad f(v) = \int_0^v \frac{v_1 \, dv_1}{g + h(v_1)} .$$

In the particular case where $h(v) = 0$, we obtain the standard result, $v^2/2g$, for the maximum height above the ground.

2.12 Vertical Motion—II

Consider the more general case where motion is through an inhomogeneous medium, or, generally, where acceleration depends upon velocity. Let the defining equation be for $u > 0$,

$$(2.32) \qquad u'' = h(u, u'), \quad u(0) = c_1, \quad u'(0) = c_2.$$

Assume that $h(u, u') \leq 0$ for all u and u', so that $c_1 = 0$, $c_2 = 0$ implies no motion.

The maximum altitude is now a function of both c_1 and c_2. Introduce the function

$$(2.33) \qquad f(c_1, c_2) = \text{the maximum altitude attained starting}$$
$$\text{with the initial position } c_1 \text{ and}$$
$$\text{initial velocity } c_2.$$

Then, as above, we obtain the approximate functional equation

$$(2.34) \qquad f(c_1, c_2) = c_2 \Delta + f(c_1 + c_2 \Delta, c_2 + h(c_1, c_2)\Delta) + O(\Delta^2).$$

This yields, in the limit, the partial differential equation

$$(2.35) \qquad c_2 + c_2 \frac{\partial f}{\partial c_1} + h(c_1, c_2) \frac{\partial f}{\partial c_2} = 0.$$

Computational aspects of problems of this nature will be discussed below, in Chapter 5. Let us merely note here that since c_2 is monotone decreasing, it can be used as a time-like variable.

2.13 Maximum Altitude

Consider now the case where motion takes place in a plane. Let the equations be

$$(2.36) \qquad x'' = g(x', y'), \quad x(0) = 0, \quad x'(0) = c_1,$$
$$y'' = h(x', y'), \quad y(0) = 0, \quad y'(0) = c_2.$$

Introducing, as before, the function $f(c_1, c_2)$, equal to the maximum altitude, we obtain the approximate functional equation

$$(2.37)$$
$$f(c_1, c_2) = (c_1^2 + c_2^2)^{1/2}\Delta + f(c_1 + g(c_1, c_2)\Delta, c_2 + h(c_1, c_2)\Delta) + O(\Delta^2).$$

47

Hence,

$$(2.38) \qquad (c_1^2 + c_2^2)^{1/2} + g(c_1, c_2)\, \frac{\partial f}{\partial c_1} + h(c_1, c_2)\, \frac{\partial f}{\partial c_2} = 0.$$

This can be treated analytically by means of the method of characteristics, and explicitly solved if g and h are of simple enough nature.

2.14 Maximum Range

To tackle the question of maximum range directly requires the introduction of another state variable, the initial altitude. We can decompose this problem into two problems, corresponding to the ascent to maximum altitude, and to the descent.

2.15 Multi-stage Processes and Differential Equations

We have seen that a physical process ruled by a differential equation, or a difference equation, gives rise to a sequence of transformations of the original state vector.

Let us now think in more general terms. Let the system S be specified by a point p in an appropriate phase space, and at a set of times $t = 0, 1, 2, \ldots$, let a transformation $T(p)$ operate upon the system. We then obtain a succession of states

$$(2.39) \qquad p_1 = T(p), \quad p_2 = T(p_2), \ldots, p_n = T(p_{n-1}), \ldots.$$

We call a process of this type a *multi-stage process*. Much of mathematical physics can be considered to be the study of particular multi-stage processes. This point of view, applied to various physical processes, with associated transformations, constitutes the kernel of the theory of invariant imbedding.

Bibliography and Discussion

2.1 The modern abstract theory of dynamical systems, inaugurated by Poincaré and extensively cultivated by Birkhoff and others since then, may be found in

G. D. Birkhoff, *Dynamical Systems*, Amer. Math. Soc. Colloq. Publ., vol. IX, 1927.

W. H. Gottschalk and G. A. Hedlund, *Topological Dynamics*, Amer. Math. Soc. Colloq. Publ., vol. XXXVI, 1955.

Many further references will be found in these volumes. It is rather interesting to note that these ideas can be applied to the question of determining whether or not an infinite chess game is possible. See

M. Morse and G. A. Hedlund, "Symbolic dynamics," *Amer. J. Math.*, vol. 60, 1938, pp. 815–866.

F. Bagemihl, "Transfinitely endless chess," *Z. fur Math., Logik Grundlagen*, vol. 2, 1956, pp. 215–217.

For the classical theory, see

E. T. Whittaker, *A Treatise on the Analytic Dynamics of Particles and Rigid Bodies*, Cambridge Univ. Press, 1944, and for an interesting extension of the ideas of Hamilton, see

J. L. Synge, *Geometrical Mechanics and de Broglie Waves*, Cambridge Univ. Press, 1954.

2.2 In connection with these ideas, it would be appropriate to discuss both the Lie theory of continuous transformation-groups and the characteristic function of Hamilton. In connection with the first named, see

E. L. Ince, *Ordinary Differential Equations*, Dover Publications, Inc., New York, 1944,

where other references may be found.

For the second, see the book by E. T. Whittaker referred to above.

2.3 For a detailed discussion of causality and its connections with functional equations of all types, see

J. Hadamard, *Lectures on Cauchy's Problem in Linear Partial Differential Equations*, Dover, 1953.

2.4 It is rather interesting to observe, as first done by Cauchy, that the functional equations, under mild assumptions, determine the functions.

2.7 For the fundamental properties of linear systems and the matrix exponential, see

R. Bellman, *Stability Theory of Differential Equations*, McGraw-Hill Book Co., Inc., New York, 1954.

It was first shown by G. Polya that the functional equation determines the matrix exponential; see

R. Bellman, *Matrix Analysis*, McGraw-Hill Book Co., Inc., New York, 1959. For an intensive discussion of semi-groups and functional equations, see

E. Hille, *Functional Analysis and Semi-groups*, Amer. Math. Soc. Colloq. Publ., vol. XXXI, 1948.

For another application of groups of transformations to determine properties of Legendre polynomials, see

E. Wigner, *Group Theory*, Academic Press, 1959.

2.8 Those interested in reading of further results in the very important domain of iteration may wish to consult

J. Hadamard, "Two works in iteration and related questions," *Bull. Amer. Math. Soc.*, vol. 50, 1944, pp. 67–75.

H. Töpfer, "Komplexe Iterationsindizes ganzer und rationales Funktionen," *Math. Annalen*, vol. 121, 1949–1950, pp. 191–222.

R. Bellman, "On the iteration of power series in two variables," *Duke Math. J.*, vol. 19, 1952, pp. 339–347,

where many further references will be found.

The theory of iteration plays an important role in the theory of branching processes; processes which include such diverse phenomena as cosmic ray cascade processes, neutron fission and biological mutation. More realistic description of these processes requires a generalization of iteration, arising from the study of age-dependent branching processes. See

R. Bellman and T. E. Harris, "On age-dependent stochastic branching processes," *Proc. Nat. Acad. Sci. USA*, vol. 34, 1948, pp. 149–152.

T. E. Harris, "Some mathematical models for branching processes," *Proc. Berkeley Symposium on Probability and Statistics*, 1952,

and a forthcoming monograph,

T. E. Harris, *Branching Processes*, Ergebnisse der Mathematik, 1960.

This theory in turn has a great deal in common with the theory of invariant imbedding,

R. Bellman and R. Kalaba, "On the principle of invariant imbedding and propagation through inhomogeneous media," *Proc. Nat. Acad. Sci. USA*, vol. 42, 1956, pp. 629–632.

—————————, "Invariant imbedding, wave propagation and the WKB approximation," *Proc. Nat. Acad. Sci. USA*, vol. 44, 1958, pp. 317–319.

R. Bellman, R. Kalaba, and G. M. Wing, "On the principle of invariant imbedding and one-dimensional neutron multiplication," *Proc. Nat. Acad. Sci. USA*, vol. 43, 1957, pp. 517–520.

The functional equations of Abel, Koenigs and others, derived from deterministic iteration, can also be generalized; see

R. Bellman, *On Limit Theorems for Non-commutative Operations—II: A Generalization of a Result of Koenigs*, The RAND Corporation, Paper P-485, 1954.

A presentation of the fundamental results of von Neumann and G. D. Birkhoff in ergodic theory may be found in the monograph

E. Hopf, *Ergodentheorie*, Engebnisse der Math., vol. 2, J. Springer, Berlin, 1937.

which contains a history of the problems in this area and many further results.

For some ergodic results established in a quite simple manner, see

R. Bellman, "An Ergodic Theorem," 1961, to appear.

2.9 This linearization was given by Carleman in connection with his work in the ergodic theorem, see

T. Carleman, "Applications de la théorie des équations intégrales singulières aux équations différentielles de la dynamique," *Ark. Mat. Astron. Fys. 22B* No. 7 (1932), pp. 1–7.

2.10–2.14 These results are contained in

R. Bellman, "Functional equations and maximum range," *Q. Appl. Math.*, to appear.

A further application of this idea may be found in

M. Ash, R. Bellman, and R. Kalaba, "Control of reactor shutdown to minimize xenon poisoning," *Nuclear Engineering*, to appear.

CHAPTER III

MULTISTAGE DECISION PROCESSES AND DYNAMIC PROGRAMMING

Are you the practical dynamic son-of-a-gun?
Have you come through with a few abstractions?

CARL SANDBURG
"Mr. Attila"

3.1 Introduction

In this chapter we wish to introduce the concept of a multistage decision process. We shall motivate and illustrate processes of this type by means of the dynamical systems and point transformations discussed in the preceding chapter. The justification of the effort put into this approach will lie in the new treatment of the calculus of variations we shall give in the following chapter, and the subsequent applications to deterministic control, stochastic control and adaptive control processes throughout the remainder of the book.

We shall first point out the type of mathematical problem which arises when multistage decision processes are treated in classical terms and explain why the analytic and computational difficulties that arise force us to devise new methods for dealing with these questions.

We turn then to the theory of dynamic programming, whose basic approach is founded upon the use of functional equations. These equations are quite similar in form to those we have already encountered in the study of dynamical systems. One difference is that the role of the principle of causality is now played by a new principle, the "principle of optimality."

Discrete deterministic processes will be discussed first, and then processes of continuous type. For this latter case we shall derive a nonlinear partial differential equation of novel type, an analogue of the equation derived in 2.5.

3.2 Multistage Decision Processes

In the preceding chapter it was pointed out that the behavior of a dynamical system over time could be interpreted in terms of the iteration

51

of a point transformation. A first extension of a process of this type is one in which the transformation that is applied is dependent upon the time at which it is applied. This situation arises when the dynamical system is governed by an equation containing time explicitly,

(3.1) $$\frac{dx}{dt} = g(x, t), \quad x(0) = c.$$

Observe that by increasing the dimensionality of x by one, and letting t be x_{N+1}, we can always write (3.1) in the form

(3.2) $$\frac{dy}{dt} = h(y), \quad y(0) = c',$$

where

(3.3) $$y = \begin{pmatrix} x_1 \\ x_2 \\ \cdot \\ \cdot \\ \cdot \\ x_N \\ x_{N+1} \end{pmatrix}.$$

It follows that no new ideas are required for the study of processes of this nature.

A second extension, one that *does* require some new ideas and one that we will fasten our attention upon, is a process in which we have a choice of the transformation to be applied at any particular time. At each stage, there will be a set of available transformations, a set which we wish to have depend not only upon the time, but also upon the current state of the system. The transformation that is applied will be dependent upon the state of the system.

A process of this type we call a *multistage decision process*. To make this notion precise, consider a family of transformations, $\{T(p, q)\}$, where q represents a vector variable which specifies a particular member of the family. We shall call q the *decision variable*. The range of q determines the set of admissible transformations. In general, the q that will be chosen will depend upon p, which is to say it will be a function of p, $q = q(p)$. A choice of q is then a choice of a transformation. We shall then use the terms *"decision"* and *"transformation"* interchangeably.

3.3 Discrete Deterministic Multistage Decision Processes

We shall begin by considering the simplest type of multistage decision process, a *discrete deterministic* process. This is a natural extension of an iteration process.

52

By a *discrete* process, we mean a process where decisions, i.e. transformations, are made at a finite, or at most, denumerable number of times. By *deterministic*, we mean that a choice of q, a decision, determines a unique outcome, $T(p, q)$, given the initial state p.

Consider now a process which unfolds in the following fashion. At an initial time and in an initial state which we shall call p_1, an initial decision, q_1, is made. The result is a new state, p_2, given by the relation

$$(3.4) \qquad p_2 = T(p_1, q_1).$$

At this point a second decision, q_2, is made, resulting in a new state, p_3, determined by the relation

$$(3.5) \qquad p_3 = T(p_2, q_2).$$

The process continues in this way with p_N for general N given by the relation

$$(3.6) \qquad p_N = T(p_{N-1}, q_{N-1}).$$

The question now arises as to what dictates the choice of the q_i. Let us answer this question for the case where we are dealing with a process involving only a finite number of decisions, say N. Occasionally we shall call this an *N-stage process*. We shall have no occasion here to discuss infinite stage processes, although we shall consider some where the duration is not fixed in advance, but depends upon the course of the process. These arise in the study of implicit variational problems. Although infinite processes are always mathematical approximations to the more realistic finite processes, they often yield useful approximate policies, and are, in many cases, simpler analytically.

Let us associate with the N-stage process a scalar function

$$(3.7) \qquad F(p_1, p_2, \ldots, p_N; q_1, q_2, \ldots, q_N),$$

which we shall use to evaluate a particular sequence of decisions q_1, $q_2, \ldots, q_N$ and states $p_1, p_2, \ldots, p_N$. This will be called the *criterion function* or the *return function*. We shall then suppose that the purpose of the decision process is to choose the q_i so as to maximize this function of the p_i and q_i.

3.4 Formulation as a Conventional Maximization Problem

Let us now see what the solution of this maximization problem involves. Since the p_i are related to the q_i by means of the relations in (3.6), and p_1 is given, it is easily seen that we can, if we wish, write

$$(3.8) \qquad F(p_1, p_2, \ldots, p_N; q_1, q_2, \ldots, q_N) = G(q_1, q_2, \ldots, q_N),$$

eliminating the dependence upon the p_i.

The problem of determining a set of q-values, $[q_1, q_2, \ldots, q_N]$, which maximizes the function $G(q_1, q_2, \ldots, q_N)$ can therefore be regarded as a particular case of the general problem of maximizing a function of N variables. This approach, however, as we hasten to point out, is definitely not a profitable one in most cases.

In the first place, we cannot apply the routine technique of setting partial derivatives equal to zero and expect immediate success. Unless G is a function of particularly simple form, the problem of solving the N simultaneous equations

$$(3.9) \qquad \frac{\partial G}{\partial q_i} = 0, \quad i = 1, 2, \ldots, N,$$

cannot be resolved analytically. Even in the case where these are linear equations, possessing an explicit analytic solution, this may still not be a feasible method of solution of the original maximization problem.

Generally, the equations in (3.9) must be solved by some type of search process. If this is the case, we may just as well use search techniques to determine the maximum of the function G.

Considering the further problems of relative maxima, constraints, and so on, we see that we face the same formidable difficulties mentioned in the introductory chapter on the calculus of variations.

Finally, we cannot, in cavalier fashion, dismiss the problem to the tender care of a computer. Numerical determination of the maximum of $G(p_1, q_1, q_2, \ldots, q_N)$ by a straightforward search technique is clearly impossible if N is large. Even in the case where we possess quite detailed information concerning the structure of G, e.g. convex or concave with a unique extremum and so on, we do not have efficient methods for locating the extremum if $N > 1$.

The solution of the most general multistage decision processes must then be regarded as impossible in terms of present day analytic and computational capabilities. We neither possess powerful analytic approaches nor feasible numerical techniques. To make any progress we must lower our sights.

3.5 Markovian-type Processes

In order to tackle this problem by the methods we shall present subsequently, we must introduce some assumptions concerning the form of the function $F(p_1, p_2, \ldots, p_N; q_1, q_2, \ldots, q_N)$. The basic property that we wish F to possess is one of *Markovian* nature, to wit:

After any number of decisions, say k, we wish the effect of the remaining $N - k$ stages of the decision process upon the total return to depend only upon the state of the system at the end of the k-th decision and the subsequent decisions.

Examples of return functions for which this property holds are

(3.10) (a) $\quad g(p_1, q_1) + g(p_2, q_2) + \ldots + g(p_N, q_N),$

(b) $\quad g(p_N, q_N)$ ("terminal control").

As we shall see, return functions of this type arise naturally in the study of control processes, in general, and particularly in the classical calculus of variations. We shall show, however, that many problems in which the criterion function cannot be expressed in analytic terms also possess this Markovian property. As a matter of fact, this property is characteristic of problems arising from engineering and economic control processes. Nonetheless, there are many important processes which cannot be unravelled in this one-dimensional fashion.

3.6 Dynamic Programming Approach

Let us consider the problem of maximizing the function

$$(3.11) \qquad g(p_1, q_1) + g(p_2, q_2) + \ldots + g(p_N, q_N)$$

over all values of the q_i, where, as before,

$$(3.12) \qquad p_i = T(p_{i-1}, q_{i-1}), \quad i = 2, 3, \ldots, N.$$

We shall proceed formally, assuming that the maximum exists. One way of obviating all rigorous details is to allow the q_i to range over only a finite set of values, and to suppose that the function $g(p, q)$ is finite for all finite p and q.

Our basic idea is not to regard this problem as an isolated problem for fixed values of p_1 and N, but instead to imbed it within a family of maximization processes. It may well be that although various members of the family are difficult to treat by themselves, there may yet exist simple relations connecting various members of the family. If then we can find one member of the family of maximization problems which has a simple solution, we can employ the relations linking various members of the family to deduce an inductive solution of the general problem.

In order to specify a family of maximization processes that can be generated by means of problems of this nature, we observe that the actual maximum value of the function appearing in (3.11) depends upon two quantities:

(3.13) (a) $\quad p_1 =$ the initial state variable,

(b) $\quad N =$ the number of stages in the multistage decision process.

Letting p_1 range over some domain of values, and N assume the values $1, 2, \ldots$, we generate a sequence of functions, $\{f_N(p_1)\}$, defined by the relation

$$(3.14) \quad f_N(p_1) = \underset{\{q_i\}}{\operatorname{Max}} \left[g(p_1, q_1) + g(p_2, q_2) + \ldots + g(p_N, q_N) \right]$$

Observe that one member of this sequence of functions is relatively easy to obtain, namely the function $f_1(p_1)$. It is given by the relation

$$(3.15) \qquad f_1(p_1) = \operatorname*{Max}_{q_1} g(p_1, q_1).$$

As a maximization only over q_1, it is the simplest of the maximization problems posed above. The question is now how to obtain relations connecting various elements of the sequence $\{f_N(p_1)\}$.

3.7 A Recurrence Relation

Let us first of all proceed analytically. Following this, we shall dissect and analyze the result we obtain. Write

$$(3.16) \qquad f_N(p_1) = \operatorname*{Max}_{q_1} \operatorname*{Max}_{q_2} \ldots \operatorname*{Max}_{q_N} [g(p_1, q_1) + g(p_2, q_2) + \ldots + g(p_N, q_N)].$$

The separability of the criterion function enables us to write this relation in the form

$$(3.17) \qquad f_N(p_1) = \operatorname*{Max}_{q_1} [g(p_1, q_1) + \operatorname*{Max}_{q_2} \operatorname*{Max}_{q_3} \ldots \operatorname*{Max}_{q_N} \{g(p_2, q_2) + \ldots + g(p_N, q_N)\}].$$

Observe the expression

$$(3.18) \qquad \operatorname*{Max}_{q_2} \operatorname*{Max}_{q_3} \ldots \operatorname*{Max}_{q_N} \{g(p_2, q_2) + \ldots + g(p_N, q_N)\},$$

for $N \geq 2$.

This represents the return from an $(N-1)$-stage decision process, starting in the state p_2. Hence, we have the relation

$$(3.19) \qquad f_{N-1}(p_2) = \operatorname*{Max}_{q_2} \operatorname*{Max}_{q_3} \ldots \operatorname*{Max}_{q_N} \{g(p_2, q_2) + \ldots + g(p_N, q_N)\}.$$

Using this result, (3.17) simplifies greatly and becomes

$$(3.20) \qquad f_N(p_1) = \operatorname*{Max}_{q_1} [g(p_1, q_1) + f_{N-1}(p_2)].$$

Since $p_2 = T(p_1, q_1)$, we may write this in the form

$$(3.21) \qquad f_N(p_1) = \operatorname*{Max}_{q_1} [g(p_1, q_1) + f_{N-1}(T(p_1, q_1))],$$

for $N = 2, 3, \ldots$. This is the desired relation connecting various members of the sequence $\{f_N(p_1)\}$.

3.8 The Principle of Optimality

Let us now see how we could have derived this result ab initio, using a very simple, yet powerful, principle. To begin with, let us introduce some new terminology. A sequence of allowable decisions, $\{q_1, q_2, \ldots, q_N\}$, will be called a *policy*; specifically, an N-stage policy. A policy which yields the maximum value of the criterion function, that is to say a policy which produces $f_N(p_1)$, will be called an *optimal policy*.

Let us suppose that we are dealing with decision processes possessing the

Markovian property described above. In this case, the basic property of optimal policies is expressed by the following:

PRINCIPLE OF OPTIMALITY. An optimal policy has the property that whatever the initial state and the initial decision are, the remaining decisions must constitute an optimal policy with regard to the state resulting from the first decision.

A proof by contradiction is immediate.

3.9 Derivation of Recurrence Relation

Using the principle cited above, we can derive the recurrence relation of (3.21) by means of purely verbal arguments. Suppose that we make an initial decision q_1. The result of this decision is to transform p_1 into $T(p_1, q_1)$ and to reduce an N-stage process to an $(N-1)$-stage process. By virtue of the principle of optimality, the contribution to the maximum return from the last $(N-1)$-stages will be $f_{N-1}(T(p_1, q_1))$. This is the Markovian property discussed in 3.5.

Hence, for some q_1 we have

(3.22) $$f_N(p_1) = g(p_1, q_1) + f_{N-1}[T(p_1, q_1)].$$

It is clear that this q_1 must be chosen to maximize the right-hand side of (3.22), with the result that the final equation is

(3.23) $$f_N(p_1) = \operatorname*{Max}_{q_1} [g(p_1, q_1) + f_{N-1}[T(p_1, q_1)]].$$

Although we shall discuss these matters in much detail in Chapter V which is devoted to the computational aspects, let us make this preliminary observation here. Observe that the usual approach to the solution of the maximization problem yields a solution in the form $[q_1, q_2, \ldots, q_N]$, the same one, of course, which is obtained as a result of our variational process. The approach we present here yields the solution in steps: first, the choice of q_1, then the choice of q_2, and so on.

The decomposition of the problem of choosing a point in N-dimensional phase space into N choices of points in one-dimensional phase space is of the utmost conceptual, analytic, and computational importance in all of our subsequent work. The whole book is essentially devoted to the study of various classes of multidimensional processes which can, by one device or another, be reduced to processes of much lower dimension, and preferably to one-dimensional processes.

3.10 "Terminal" Control

A case of some interest in many significant applications is that where we wish only to maximize a function of the final state p_N. In this case, the sequence of return functions is defined by the relation

(3.24) $$f_N(p_1) = \operatorname*{Max}_{q_i} g(p_N).$$

Since it is clear that this process possesses the requisite Markovian property, we see that the basic recurrence relation is now

$$(3.25) \qquad f_N(p_1) = \text{Max}\, f_{N-1}(T(p_1, q_1)), \quad N = 2, 3, \ldots ,$$
$$f_1(p_1) = g(p_1).$$

3.11 Continuous Deterministic Processes

To define a continuous decision process requires some care because of a number of subtle conceptual difficulties that arise whenever we pass from the finite to the infinite. As we shall see, the calculus of variations furnishes a natural and important class of continuous decision processes. Let us, proceeding formally, derive the concept of a continuous decision process from that of a discrete decision process. There are, however, direct ways of formulating these processes.

Let decisions be made at the times $0, \Delta, 2\Delta, \ldots$, and so on, and introduce the "rate of return," the function $g(p_1, q_1)\Delta$. Since Δ is to be considered to be an infinitesimal we must write $T(p_1, q_1)$, the transformation induced by the decision q_1, in the form

$$(3.26) \qquad T(p_1, q_1) = p_1 + \Delta S(p_1, q_1) + O(\Delta^2).$$

Let us write $t = N\Delta$, and $f_N(p_1) = f(p_1, t)$. Then (3.23) takes the form

$$(3.27) \quad f(p_1, t) = \underset{q_1}{\text{Max}}\, [g(p_1, q_1)\Delta + f[p_1 + \Delta S(p_1, q_1), t - \Delta]] + O(\Delta^2).$$

Expanding in powers of Δ, we have, formally,*

$$(3.28) \quad f[p_1 + \Delta S(p_1, q_1), t - \Delta] = f(p_1, t) - \Delta \frac{\partial f}{\partial t}(p_1, t)$$
$$+ \Delta\left(S(p_1, q_1), \frac{\partial f}{\partial p_1}\right) + O(\Delta^2).$$

Here (x, y) denotes the inner product, and

$$(3.29) \qquad \frac{\partial f}{\partial p_1} = \begin{pmatrix} \dfrac{\partial f}{\partial p_{11}} \\[1em] \dfrac{\partial f}{\partial p_{12}} \\ \cdot \\ \cdot \\ \cdot \\ \dfrac{\partial f}{\partial p_{1N}} \end{pmatrix},$$

---

* By the term "formally" we mean without attention to the rigorous aspects such as the existence of partial derivatives, interchange of limits, and so on. It is a fine word to use to cover a multitude of mathematical sins.

where p_{1j}, $j = 1, 2, \ldots, M$, are the components of the M-dimensional vector p_1.

Passing to the limit as $\Delta \to 0$, we obtain the nonlinear partial differential equation

$$(3.30) \qquad \frac{\partial f}{\partial t} = \operatorname*{Max}_{q_1} \left[g(p_1, q_1) + \left[S(p_1, q_1), \frac{\partial f}{\partial p_1} \right] \right].$$

We may either use this equation to *define* the concept of a continuous decision process, or, in the case where other formulations exist, as, for example, the calculus of variations, derive this equation. It is the analogue for variational processes of the equation of 2.5. In the next chapter, we shall examine these matters in detail.

3.12 Discussion

The fundamental device that we have employed is that of imbedding a particular process within a family of similar processes. An essential point to emphasize is that there can be many ways of accomplishing this. Sometimes, it is clear how to do this; sometimes a certain amount of ingenuity is required.

As we shall see in the chapter on computational aspects, in all cases we face the problems of feasible and efficient applications of these general ideas. We constantly try to decompose complicated processes into sequences of simple processes. From this point of view dynamic programming is not so much any fixed set of analytic techniques as a state of mind.

Bibliography and Comments

We follow here the discussion given principally in Chapter 3 of

R. Bellman, *Dynamic Programming*, Princeton Univ. Press, Princeton, N.J., 1957.

There is really no clear-cut difference between descriptive and variational processes, since most of the important physical processes can be cast as control processes by means of a variety of variational principles.

3.4 For a discussion of the method of S. Johnson to determine optimal search techniques for finding the maximum of a function of one variable, see page 34 of the above cited book. See also

O. Gross and S. Johnson, "Sequential minimax search for a zero of a convex function," *Math. Tables and other Aids to Computation*, vol. XIII, 1959, pp. 44–51.

D. J. Newman, "Locating the maximum point on a unimodal surface," *Notices of the Amer. Math. Soc.*, vol. 6 (1959), p. 799.

and the paper by J. Kiefer referred to in the Bibliography of 16.1.

3.5 Markov processes are discussed in some detail in Chapter IX and Markovian decision processes in Chapter XI.

3.9 In certain cases, we can obtain formulas completely analogous to those holding in the study of descriptive processes. Consider, for example, the problem of determining the maximum of the function $\sum_{i=1}^{N} g(x_i)$ over the region determined by $\sum_{i=1}^{N} x_i = c$, $x_i \geq 0$. Then, with the usual notation, we have

$$f_{M+N}(c) = \text{Max} \left[\sum_{i=1}^{M} g(x_i) + \sum_{i=M+1}^{M+N} g(x_i) \right],$$

whence

$$f_{M+N}(c) = \underset{0 \leq c_1 \leq c}{\text{Max}} \, [f_M(c_1) + f_N(c - c_1)],$$

for $M, N \geq 1$.

This result can be used to effect a great saving of computational effort if one wishes to solve a particular problem, say that of determining $f_{1024}(c)$. In place of computing all the members of the sequence $\{f_k(c)\}$, $k = 1, 2, \ldots, 1024$, we need only tabulate $f_1(c), f_2(c), f_4(c), f_8(c)$, and so on, which is to say $\{f_{2^k}(c)\}$, for $k = 0, 1, \ldots, 10$.

3.10 See, for a different treatment and further references,

R. Radner, Paths of Economic Growth That are Optimal With Regard Only to Final States; Two "Turnpike Theorems," Bureau of Business and Economic Research, Univ. of Calif., Berkeley, 1960.

CHAPTER IV

DYNAMIC PROGRAMMING AND THE
CALCULUS OF VARIATIONS

Sweet are the uses of Adversity.

WILLIAM SHAKESPEARE
As You Like It, Act II, Scene I.

4.1 Introduction

In this chapter, we shall use the techniques developed in the foregoing chapter to treat problems in the calculus of variations. We shall consider only questions involving one dependent variable, although the same methods are applicable to the general case involving any finite number of dependent variables. Problems involving functions of two or more variables require, on the other hand, more sophisticated concepts.

We will require no previous knowledge of the calculus of variations on the part of the reader. All results required for the computational solution will be derived *ab initio*.

Our first task will be to show that the problem of minimizing an integral such as

$$(4.1) \qquad J(u) = \int_0^T g(u, u')\, dt,$$

or, more generally, minimizing the integral

$$(4.2) \qquad J(y) = \int_0^T g(u, v)\, dt$$

over all functions v, where u and v are connected by means of the differential equation

$$(4.3) \qquad \frac{du}{dt} = h(u, v), \quad u(0) = c,$$

may be construed to be a multistage decision process of continuous type. Once we have established this, the application of the functional equation technique introduced in the previous chapter is fairly routine.

After treating the continuous version to the degree of deriving the Euler equation, we shall turn to the discrete version. As we shall see in the following chapter, this discrete version will be the foundation of our theory of numerical solution of variational problems.

We shall indicate, briefly, how these techniques are applicable to the study of trajectory problems and to the "bang-bang" control problem.

Finally, we shall discuss the concept of *sequential computation* which is intimately related to the general study of adaptive control processes, and thus provides a hint of what is to follow.

4.2 The Calculus of Variations as a Multistage Decision Process

The problem of determining a function u which minimizes an integral such as $\int_0^T g(u, u')\, dt$ may, of course, be considered to be a problem of determining the function u'. Put in these terms, we see that at every point on the extremal curve $u(t)$ we determine a direction along which the path is to be continued.

In place of prescribing u as a function of t, we could just as well prescribe u' as a function of t, or, more importantly from our point of view, as a function of u and t. In other words, regarding a choice of u' as a choice of a policy, we can either describe the process *in toto*, or we can describe the decision to be made in terms of the state of the system at the time the decision is made.

For the case of deterministic processes we have a choice as to which description we shall use. For the case of stochastic processes which we shall discuss subsequently, we have no choice. We must employ the state variable description.

It is in this way, by regarding the determination of $u(t)$ as equivalent to the determination of $u'(t)$ at any point, that we can consider the variational problems described above as multistage decision processes of continuous type.

We wish then to indicate how the calculus of variations can be treated by an extension of the semi-group concept.

4.3 Geometric Interpretation

The foregoing comments can be interpreted very elegantly in geometric terms. The classical approach to the calculus of variations regards a curve as a *locus of points*—the dynamic programming approach regards a curve as an *envelope of tangents*.

The equivalence of these concepts is the fundamental duality property of the two-dimensional Euclidean plane. It is to be expected that an approach to the calculus of variations which exploits the full geometric properties will be more powerful than one which is based solely on one approach or the other.

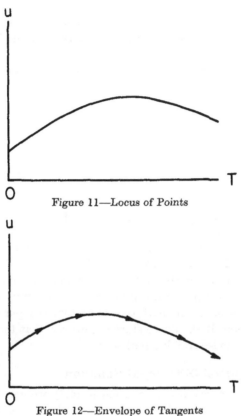

Figure 11—Locus of Points

Figure 12—Envelope of Tangents

4.4 Functional Equations

Let us now put all these ideas together to obtain a treatment of the calculus of variations by means of functional equations. Consistent with our general aims, we are interested only in laying clear the basic ideas and will not stop to fill in the rigorous details, which, unfortunately or not, are not trivial.

Consider, initially, the problem of minimizing

$$(4.4) \qquad J(u) = \int_0^T g(u, u') \, dt,$$

over all functions $u(t)$ satisfying the initial condition $u(0) = c_1$.

Let us then introduce the following function of two variables

$$(4.5) \qquad f(c, T) = \underset{u}{\text{Min}} \int_0^T g(u, u') \, dt.$$

Observe that we are imbedding the original problem, in which c and T were fixed quantities, within a family of processes in which c and T are parameters.

The minimum is taken over all functions satisfying the condition $u(0) = c$. The variable c can assume any real value, while T is constrained to be non-negative.

A choice of $u'(t)$ over the interval $[0, T]$ consists of a choice over $[0, S]$ plus a choice over $[S, T]$. As a result of a choice of $u'(t)$ over $[0, S]$, the initial value c is transformed into

$$(4.6) \qquad c' = c + \int_0^S u'(t) \, dt.$$

From the additive character of the integral, (which yields the desired Markovian property discussed in 3.5) plus the principle of optimality, we deduce the functional equation

$$(4.7) \quad f(c, T) = \operatorname*{Min}_{u'[0,S]} \left[\int_0^S g(u, u') \, dt + f\left[c + \int_0^S u'(t) \, dt, \, T - S \right] \right].$$

The minimum is to be taken over all functions $u'(t)$ defined over $[0, S]$.

As will be noted, we have not stopped to impose conditions of various types which will ensure existence and uniqueness of solutions of the variational problems. Rather, we are proceeding in a purely formal fashion, completely analogous to the path that is customarily pursued in deriving the Euler equation. It should be stated, however, that the discussion can be made rigorous in several different ways.

4.5 Limiting Partial Differential Equation

In order to make use of the equation in (4.7), we consider S to be an infinitesimal. Then choice of u' over an interval comes down to choice of u' at $t = 0$, an initial direction.

Let

$$(4.8) \qquad v = v(c, T) = u'(0),$$

indicating that the optimal initial slope is a function of the initial state c and the duration of the process T. The choice of v as a function of c and T determines a policy. Then, expanding around $S = 0$, we have the relations

$$(4.9) \quad (a) \quad \int_0^S g(u, u') \, dt = g(c, v)S + O(S^2),$$

$$(b) \quad c + \int_0^S u'(t) \, dt = c + Sv + O(S^2),$$

$$(c) \quad f\left[c + \int_0^S u'(t) \, dt, \, T - S \right] = f[c + Sv, \, T - S] + O(S^2)$$

$$= f(c, T) + Sv \frac{\partial f}{\partial c} - S \frac{\partial f}{\partial T} + O(S^2).$$

Thus, (4.7) reduces to the approximate equation

$$(4.10) \quad f(c,\,T) = \underset{v}{\text{Min}} \left[Sg(c,\,v) + f(c,\,T) + Sv\,\frac{\partial f}{\partial c} - S\,\frac{\partial f}{\partial T} \right] + O(S^2).$$

Cancelling the terms $f(c,\,T)$ and dividing through by S, we obtain the limiting form

$$(4.11) \qquad\qquad \frac{\partial f}{\partial T} = \underset{v}{\text{Min}} \left[g(c,\,v) + v\,\frac{\partial f}{\partial c} \right],$$

a special case of (3.30). The initial condition is clearly

$$(4.12) \qquad\qquad\qquad f(c,\,0) = 0.$$

As we shall show below, this new equation contains the classical Euler equation. What is perhaps more important is that we will be able to use (4.11), or discrete versions, to handle problems in which no Euler equations need exist.

4.6 The Euler Equations and Characteristics

Still proceeding formally, let us note that (4.11) is in reality equivalent to two equations,

$$(4.13) \qquad\qquad\qquad \frac{\partial g}{\partial v} + \frac{\partial f}{\partial c} = 0,$$

expressing the fact that v minimizes, and

$$(4.14) \qquad\qquad\qquad \frac{\partial f}{\partial T} = g(c,\,v) + v\,\frac{\partial f}{\partial c},$$

for the value of v determined by (4.8).

Since $v(c,\,T)$ is the essential function, the *policy-function*, let us obtain an equation for it by eliminating f between the two equations in (4.13) and (4.14). Using the value for $\partial f/\partial c$ obtained from (4.13) in (4.14), we obtain the two equations

$$(4.15) \quad (a) \qquad\qquad \frac{\partial f}{\partial T} = g - vg_v,$$

$$(b) \qquad\qquad\qquad \frac{\partial f}{\partial c} = -g_v.$$

Taking the partial derivative of the first relation with respect to c and the second with respect to T, and equating the two values of $\partial^2 f/\partial c\,\partial T$ obtained in this way, we end up with the partial differential equation

$$(4.16) \qquad\qquad \frac{\partial}{\partial c}(g - vg_v) = -\frac{\partial}{\partial T}(g_v)$$

or

$$(4.17) \qquad g_c + g_v v_c - v_c g_v - vg_{vv} v_c - vg_{vc} = -g_{vv} v_T,$$

for the policy-function. This yields the quasilinear partial differential equation

(4.18) $$g_{vv}v_T - vg_{vv}v_c + g_c - vg_{vc} = 0.$$

As we know, the solution of a quasilinear equation of this nature can be obtained in terms of its *characteristics*, a set of ordinary differential equations derived from the coefficient functions of (4.18). As is to be expected, these differential equations are equivalent to the Euler equation associated with the original variational problem.

The reader familiar with the theory of characteristics is urged to carry out the details. We shall present another derivation below.

4.7 Discussion

Thinking back to the duality between the approach presented above and the classical approach, and to the connection between the quasilinear partial differential equation and its characteristics, it is clear, in advance, that the characteristics must be equivalent to the Euler equation.

The fact that the relations in (4.15) bear a resemblance to the Hamilton-Jacobi equations is no coincidence, since the two theories are intimately related. The classical approach corresponds to Fermat's principle in optics, and the dynamic programming approach to Huygen's principle.

4.8 Direct Derivation of Euler Equation

Let us derive the Euler equation from (4.11). We shall obtain the more general equation corresponding to the problem of minimizing

(4.19) $$J(u) = \int_0^T g(u, u', s)\, ds.$$

Let the upper limit T be considered to be fixed, and introduce a variable lower limit t. Write

(4.20) $$f(c, t) = \operatorname*{Min}_u J(u) = \operatorname*{Min}_u \int_t^T g(u, u', s)\, ds.$$

Then, as above, we see that

(4.21) $$-\frac{\partial f}{\partial t} = \operatorname*{Min}_v \left[g(c, v, t) + v\, \frac{\partial f}{\partial c} \right].$$

This yields the equations

(4.22) $$g_v + f_c = 0,$$
$$g + vf_c = -f_t.$$

Let us now identify c with u, and v with u'. The equations are now

(4.23) $$g_{u'} + f_u = 0,$$
$$g + u'f_u = -f_t.$$

Differentiating the first with respect to t and the second with respect to u, we obtain the relations

(4.24)
$$\frac{d}{dt}\,(g_{u'}) + f_{ut} + f_{uu}u' = 0,$$

$$g_u + u'f_{uu} + f_{tu} = 0.$$

Equating f_{ut} and f_{tu}, we obtain the equation

(4.25)
$$\frac{d}{dt}\,(g_{u'}) = g_u,$$

which is precisely the Euler equation.

4.9 Discussion

In this way one can also derive a number of the classical conditions of the calculus of variations, the Legendre condition, the Weierstrass condition (for the multidimensional case as well), and treat various isoperimetric problems. Here the Lagrange multiplier is introduced, a subject we shall discuss later in great detail.

In Chapter XIX, we shall show how (4.21) can be applied to yield simple analytic relations in the case where $g(c, v, t)$ is quadratic in c and v. What is interesting to note is that in this case in which the classical techniques yield *linear*, and hence explicitly solvable, equations, the dynamic programming approach leading to nonlinear equations may still yield a more rapid computational algorithm.

This furnishes an example of the well known phenomenon that an explicit solution of a problem may not yield the best computational algorithm. The best known example of this is the Cramer solution of a system of linear equations.

4.10 Discrete Processes

As we shall see in what follows, we are less interested in continuous variational processes than we are in certain discrete versions of these processes.

Let us note in passing that it is often assumed that the discrete version is an approximation to the continuous version. Many times it is far more appropriate to consider the continuous case as a mathematical fiction employed to simplify the analysis of the actual discrete process. This is certainly the case in many control processes.

Frequently, a process which is essentially discrete in nature is formulated in continuous terms for analytic purposes. When computational results are desired, the equations describing the process, ordinary or partial differential equations, are replaced by equations describing a discrete process. Generally, this discrete process introduced for numerical purposes has nothing to do with the original discrete process.

Sometimes better results are obtained in this way. In general, however, many additional and extraneous difficulties can be introduced in this fashion.

Let us now describe a discrete version of the continuous variational problem posed in the foregoing sections. In place of allowing a choice of a direction at each point in time, let us assume that these choices can only be made at a sequence of times $t = 0, \Delta, 2\Delta, \ldots$. In place of the functions $x(t)$ and $y(t)$, we consider the sequences $\{x_k\}$, $\{y_k\}$, where $x_k = x(k\Delta)$, $y_k = y(k\Delta)$. Instead of the differential equation of (4.3), we use the difference equation

$$(4.26) \qquad x_{k+1} = x_k + g(x_k, y_k)\Delta, \quad x_0 = c,$$

and rather than an integral, we have the sum

$$(4.27) \qquad J_N(y) = \sum_{k=0}^{N-1} g(x_k, y_k)\Delta.$$

Under the sole assumption of continuity, with some boundedness restrictions on the functions g and h, we can now pose the problem of maximizing $J_N(y)$.

In order for the concept of a discrete approximation to a continuous process to be useful, it is necessary to show that the limit of the solution of the discrete process exists as $\Delta \to 0$, and in some sense approaches that of the continuous process. Treatments of this question under various assumptions and by different techniques will be found in references given at the end of the chapter.

It should be noted that there is no necessity to take the time intervals between decisions to be constant. As we shall indicate subsequently, an interesting problem in itself is that of determining what the best discrete approximations are.

4.11 Functional Equations

To treat the discrete process, introduce the sequence of functions

$$(4.28) \qquad f_N(c) = \operatorname*{Max}_{\{y\}} J_N(y),$$

for $-\infty < c < \infty$, $N = 1, 2, \ldots$. Then

$$(4.29) \qquad f_1(c) = \operatorname*{Max}_{y_0} [g(c, y_0)\Delta],$$

and, using familiar reasoning,

$$(4.30) \qquad f_{N+1}(c) = \operatorname*{Max}_{y_0} [g(c, y_0)\Delta + f_N(c + h(c, y_0)\Delta)],$$

for $N = 1, 2, \ldots$.

This equation will be the basis of our computational treatment of variational problems. We will discuss the computational aspects in the following chapter.

At the moment, let us emphasize the fact that generally the maximum in (4.30) will be determined by a search process, not by means of calculus. It is this which enables us to overcome the many difficulties involved in the usual approach of the calculus of variations.

4.12 Minimum of Maximum Deviation

Let us now employ the foregoing techniques to treat a problem which defies the usual variational approach. Using the recurrence relations for 4.10, let us determine the y_i so as to minimize the functional

$$(4.31) \qquad J_N(y) = \underset{0 \leq i \leq N-1}{\text{Max}} |x_i|.$$

To treat variational problems involving sums, we use the functional equation of addition. If

$$(4.32) \qquad T(c_1, c_2, \ldots, c_N) = c_1 + c_2 + \ldots + c_N$$

for $N = 1, 2, \ldots$, and all c_i, then it is clear that

$$(4.33) \qquad T(c_1, c_2, \ldots, c_N) = T[c_1, T(c_2, c_3, \ldots, c_N)].$$

Similarly, if

$$(4.34) \qquad M(c_1, c_2, \ldots, c_N) = \text{Max} (c_1, c_2, \ldots, c_N),$$

it is easily seen that

$$(4.35) \qquad M(c_1, c_2, \ldots, c_N) = M[c_1, M(c_2, c_3, \ldots, c_N)].$$

It is this functional property which we exploit below.

Let

$$(4.36) \qquad f_N(c) = \underset{y}{\text{Min}} J_N(y).$$

Then

$$(4.37) \qquad f_1(c) = |c|,$$

and

$$(4.38) \qquad f_N(c) = \text{Max} [|c|, \underset{y_0}{\text{Min}} f_{N-1}(c + g(c, y_0)\Delta)]$$

for $N = 2, 3, \ldots$. This is still not attractive from the analytic point of view, but it does yield a feasible computational algorithm.

4.13 Constraints

If we add the condition that

$$(4.39) \qquad |y_i| \le m, \quad i = 0, 1, \dots, (N-1),$$

then (4.38) is replaced by the relation

$$(4.40) \qquad f_N(c) = \text{Max} \, [|c|, \underset{|v_0| \le m}{\text{Min}} \, f_{N-1}(c + g(c, y_0)\Delta)].$$

Similarly, the basic partial differential equation of (4.11) is replaced by

$$(4.41) \qquad \frac{\partial f}{\partial T} = \underset{|v| \le m}{\text{Min}} \left[g(c, v) + v \frac{\partial f}{\partial c} \right].$$

As we shall indicate in the following chapter, the presence of the constraint actually simplifies the computation of the solution—when done along these lines. The analytic study is, however, complicated to an extraordinary degree.

4.14 Structure of Optimal Policy

As we have previously stated, the explicit solution of variational problems subject to constraints is usually a matter of some difficulty. Let us now indicate how the functional equation in (4.41) can be used to ascertain the structure of the optimal policy under certain reasonable assumptions concerning the function $g(c, v)$.

Once the structure has been guessed, one way or another, it is, generally speaking, not too difficult to verify in a rigorous fashion that the solution is actually of the specified form.

Let us suppose that $g(c, v)$ is a convex function of v for all c. In particular, this is the case when $g(c, v) = v^2 + h(c)$, a case of some importance. Then, regardless of the value of $\partial f/\partial c$, the function $\phi(v) = g(c, v) + v \, \partial f/\partial c$ is convex as a function of v. It follows that this function possesses a unique minimum, which may, however, occur at $v = -m$, $v = m$, or inside the interval $[-m, m]$. For any particular set of c and T values, we may be in one of the three following situations (see p. 71).

Assuming that $\phi(v)$ is continuous as a function of T, it follows from this that a T-interval where $v = m$ is optimal must be followed by an Euler regime, i.e. one in which $-m < v < m$, which can be followed by $v = +m$ or $v = -m$. However, it is impossible to have a solution consisting only of intervals in which $v = \pm m$. These transitions from one type of policy to another can occur arbitrarily often, with suitable choice of $g(c, v)$.

Suppose that we know that the function $\partial f/\partial c$ is monotone increasing as a function of T. The physical significance of this condition is clear. Then,

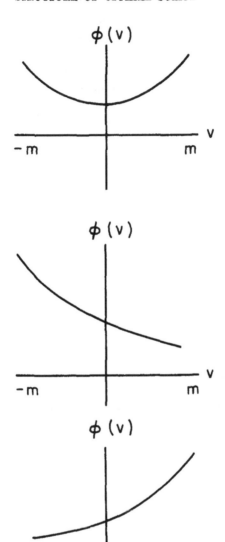

Figure 13

considering the family of curves generated by $\phi(v)$ as T increases, we see that the only possible form for an optimal policy over $[0, T_0]$ is

$$(4.42) \qquad \begin{aligned} (a) &\quad v = -m, \quad T_1 \leq T \leq T_0, \\ (b) &\quad -m < v < m, \quad T_2 \leq T \leq T_1, \\ (c) &\quad v = m, \quad 0 \leq T \leq T_2. \end{aligned}$$

71

For particular values of T, T_1 may be zero or T_1, and T_2 may be 0. All possibilities can occur, and examples of this may be found in the papers cited at the end of the chapter.

This type of analysis is particularly easy to carry out for the important case in which we wish to minimize a functional of the form

$$(4.43) \qquad J(y) = \int_0^T [(x, Bx) + \lambda(y, y)] \, dt$$

over y, where

$$(4.44) \qquad \frac{dx}{dt} = Ax + y, \quad x(0) = c,$$

and

$$(4.45) \qquad |y_i| \leq m_i, \quad i = 1, 2, \ldots, N.$$

4.15 Bang-bang Control

As an indication of the manner in which the functional equation technique can be used to treat implicit variational problems, let us consider what is often called a "bang-bang" control problem. This problem has been discussed previously in 1.18.

Let it be required to choose the function $v(t)$, subject to the constraint

$$(4.46) \qquad |v(t)| \leq m, \quad t \geq 0,$$

so as to drive the solution of

$$(4.47) \qquad u'' = g(u, u', v), \quad u(0) = c_1, \quad u'(0) = c_2,$$

to the equilibrium state, $u = u' = 0$, as rapidly as possible.

Consider the equivalent first-order system

$$(4.48) \qquad u_1' = u_2, \quad u_1(0) = c_1,$$
$$u_2' = g(u_1, u_2, v), \quad u_2(0) = c_2,$$

and the corresponding discrete version

$$(4.49) \quad u_1(n + 1) = u_1(n) + u_2(n)\Delta, \quad u_1(0) = c_1,$$
$$u_2(n + 1) = u_2(n) + g[u_1(n), u_2(n), v(n)]\Delta, \quad u_2(0) = c_2.$$

In place of requiring that u_1 and u_2 be simultaneously zero, it is sufficient to require that $u_1^2 + u_2^2 \leq \delta$, for some small value of δ.

The minimum number of stages required to force the system into the region $u_1^2 + u_2^2 \leq \delta$ in phase space will be a function only of the initial state, (c_1, c_2). Denote this function by $f(c_1, c_2)$. It follows from the principle of optimality that it satisfies the functional equation

$$(4.50) \qquad f(c_1, c_2) = 1 + \underset{|v(0)| \leq m}{\text{Min}} \, f[c_1 + c_2\Delta, c_2 + g(c_1, c_2, v)\Delta].$$

In the following chapter, we shall discuss various ways in which this equation can be utilized for computational purposes.

For the case where $g(u, u', v)$ is linear,

$$(4.51) \qquad g(u, u', v) = a_1 u + a_2 u' + v,$$

the problem can greatly be simplified by a combination of classical analysis and the functional equation. Generally, considering the linear vector equation

$$(4.52) \qquad \frac{dx}{dt} = Ax + v, \quad x(0) = c,$$

and asking the same question, one can show in many cases that the conditions

$$(4.53) \qquad |v_i(t)| \le m_i, \quad t \ge 0, \quad i = 1, 2, \ldots, N,$$

can be replaced by the constraints

$$(4.54) \qquad v_i(t) = \pm m_i, \quad t \ge 0, \quad i = 1, 2, \ldots, N.$$

The problem is extremely interesting, since its solution requires a judicious blend of classical differential equation theory and the modern theory of abstract spaces.

4.16 Optimal Trajectory

The current interest in satellites and space travel has revived the study of a number of classical trajectory problems. Generally, these problems have the following format: "Given certain allowable motions, how do we travel from one point to another in minimum time, or at minimum cost of fuel."

It is worthwhile to phrase the problem in more general terms, since it turns out that these generalized trajectory problems occur throughout the engineering world. The advantage of a simple, abstract mathematical framework lies in the fact that it furnishes us a uniform method for formulating and resolving a variety of control processes which are superficially quite different.

Let us then pose the problem in the following terms. We wish to proceed from a point p in phase space to a point p' in phase space at minimum cost in "resources," given the information that the cost of going from p to q is given by the expression $t(p, q)$.

Let the minimum cost be designated by the function $f(p)$. Then the principle of optimality yields the functional equation

$$(4.55) \qquad f(p) = \underset{q}{\text{Min}} \, [t(p, q) + f(q)].$$

This basic equation can be used to study optimal trajectories, satellite launching, geodesics, paths of minimum length through discrete networks, r-th shortest paths through networks, soft landing on the moon and other planets, chemical process control, reactor control, and many other processes and problems.

Questions of computational solution will be discussed in the next chapter.

4.17 The Brachistochrone

As an example of a problem in the calculus of variations which can be explicitly resolved, either by classical techniques, or by means of the functional equation technique which we shall pursue here, let us consider the problem of determining a curve connecting the two points P and Q with the property that a particle sliding along this curve under the sole influence of gravity will arrive at Q in a minimum time.

Without loss of generality, let P be the origin, and let the axes be oriented as indicated below.

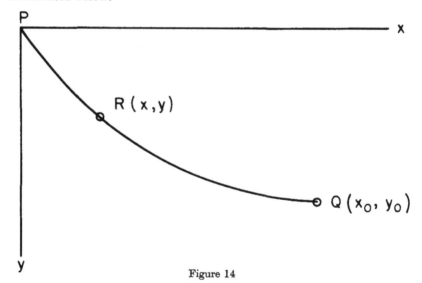

Figure 14

Omitting the gravitational constant, the problem is that of minimizing the functional

(4.56)
$$J(y) = \int_0^{x_0} \left(\frac{1 + y'^2}{y} \right)^{1/2} dx,$$

over all curves satisfying the end conditions

(4.57)
$$y(0) = 0, \quad y(x_0) = y_0.$$

74

To treat the problem by dynamic programming techniques, we let $f(x, y)$ denote the minimum time required to go from the generic point $R(x, y)$ to (x_0, y_0). The equation in (4.56) is in this case

(4.58) $\quad f(x, y) = \operatorname*{Min}_{y'} \left[\left(\frac{1 + y'^2}{y}\right)^{1/2} \Delta + f(x + \Delta, y + y'\Delta) + O(\Delta^2) \right].$ *

Passing to the limit as $\Delta \to 0$, this yields the nonlinear partial differential equation

(4.59) $\qquad 0 = \operatorname*{Min}_{y'} \left[\left(\frac{1 + y'^2}{y}\right)^{1/2} + f_x + y'f_y \right],$

equivalent to the two equations

(4.60) $\qquad (a) \quad 0 = \left(\frac{1 + y'^2}{y}\right)^{1/2} + f_x + y'f_y,$

$\qquad (b) \quad 0 = \dfrac{y'}{[y(1 + y'^2)]^{1/2}} + f_y.$

To eliminate the function f, and thus obtain an equation for y', we differentiate (4.60b) with respect to x and take the partial derivative of (4.60a) with respect to y, obtaining

(4.61) $\quad (a) \quad \dfrac{d}{dx}\left[\dfrac{y'}{(y(1 + y'^2))^{1/2}}\right] + f_{yx} + f_{yy}y' = 0,$

$\qquad (b) \quad -\tfrac{1}{2}(1 + y'^2)^{1/2}y^{-3/2} + f_{xy} + f_{yy}y' = 0.$

Thus

(4.62) $\qquad \dfrac{d}{dx}\left[\dfrac{y'}{(y(1 + y'^2))^{1/2}}\right] + \tfrac{1}{2}(1 + y'^2)y^{-3/2} = 0.$

A first integral is

(4.63) $\qquad \left(\dfrac{1 + y'^2}{y}\right)^{1/2} - \dfrac{y'^2}{[(1 + y'^2)y]^{1/2}} = k,$

or

(4.64) $\qquad \dfrac{1}{[y(1 + y'^2)^{1/2}]} = k.$

This is equivalent to Snell's law for the propagation of light. It is interesting to interpret this well-known physical principle as an optimal policy in a multistage decision process.

The integration can now be readily completed to obtain the standard representation of a brachistochrone in parameteric form:

(4.65) $\qquad y = (1 - \cos t)/2k^2,$

$\qquad x = c_1 + (t - \sin t)/2k^2.$

---

* We have deliberately kept y' as the slope, instead of v, in order to manipulate the Euler equation.

4.18 Numerical Computation of Solutions of Differential Equations

To obtain the numerical solution of a particular first order differential equation

(4.66) $$u' = g(u, t), \quad u(0) = c_1,$$

we either use an explicit analytic solution (rarely), or an analogue computer if the equation has a certain simple form (rarely), or, most commonly, we employ a digital computer.

To do this latter, as first discussed in 1.15, we must replace the transcendental operations of differentiation and integration, which involve limiting, and hence infinite, operations, by approximate processes involving addition and multiplication, and, of course, the inverse processes of subtraction and division.

One way to do this is, as outlined above, to replace the differential equation in (4.66), valid for $t \geq 0$, by the difference equation

(4.67) $$u_{k+1} - u_k = g(u_k, t_k)\Delta, \quad u_0 = c_1,$$

valid for $t = 0, \Delta, 2\Delta, \dots,$ where

(4.68) $$u_k = u(k\Delta), \quad t_k = k\Delta.$$

Expanding $u_{k+1} = u(k\Delta + \Delta)$ we have

(4.69) $$u_{k+1} = u_k + \Delta u'(k\Delta) + \frac{\Delta^2}{2} u''(k\Delta) + \dots.$$

We see then that the error in using (4.67) to compute $u[(k+1)\Delta]$ given $u(k\Delta)$ is

(4.70) $$O(\Delta^2|u''(k\Delta)|),$$

assuming that the higher order terms in (4.69) are of smaller order.

To do better at the cost of small additional effort, we can use, in place of (4.67), the recurrence relation

(4.71) $$u_{k+1} - u_{k-1} = 2g(u_k, t_k)\Delta, \quad u_0 = c_1.$$

Expanding u_{k+1} and u_{k-1} we obtain

(4.72) $$u_{k+1} = u_k + u'(k\Delta)\Delta + u''(k\Delta)\frac{\Delta^2}{2} + u'''(k\Delta)\frac{\Delta^3}{6} + \dots,$$

$$u_{k-1} = u_k - u'(k\Delta)\Delta + u''(k\Delta)\frac{\Delta^2}{2} - u'''(k\Delta)\frac{\Delta^3}{6} + \dots.$$

Hence,

(4.73) $$u_{k+1} - u_{k-1} = 2u'(k\Delta)\Delta + O[\Delta^3|u'''(k\Delta)|]$$
$$= 2g(u_k, t_k)\Delta + O[\Delta^3|u'''(k\Delta)|].$$

The recurrence relation in (4.71) is thus one order of magnitude better than that in (4.67) as far as the error term is concerned. Similarly, by using more complicated recurrence relations, we can obtain arbitrarily accurate approximations, provided that the derivatives of the solution never become unduly large. This, however, is the case in a large number of quite interesting physical processes involving shock waves, boundary layers, and so forth, where rapid changes can occur in very short times or in very small regions. For example, a shock wave is a mathematical idealization of a situation which actually looks somewhat like this:

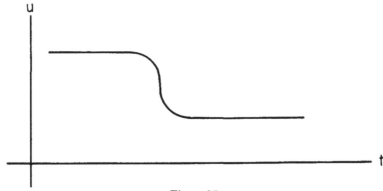

Figure 15

In this situation, first and higher derivatives assume very large values in a small t-interval. The usual approach to this problem is to use one value of Δ for the smooth portions of the curve, and another much smaller value for the part that is changing so rapidly. The practical difficulty in applying this method arises from the fact that although we usually know in advance that this quasi-shock will occur, we occasionally do not know exactly *where* it will occur. Consequently, we must watch the computational results as they come off the machine. In these days of fully automatic operations, this is a confession of weakness.

4.19 Sequential Computation

We can overcome this difficulty to some extent by using a digital computer as an adaptive device. In place of using a fixed computational technique, we shall employ a computational technique which is dependent upon the particular problem the machine is solving. We shall discuss problems of this general nature in much greater detail in subsequent chapters devoted to the study of adaptive control processes.

The way we shall employ the general method here is to introduce a variable interval, Δ_k, in place of the fixed interval, Δ, with Δ_k dependent upon the behavior of the solution at t_k.

In place of the recurrence relation of (4.67) or (4.71), we consider the two recurrence relations

(4.74)
$$u_{k+1} - u_k = g(u_k, t_k)\Delta_k, \quad u_0 = c_1,$$
$$t_{k+1} - t_k = \Delta_k.$$

Here $u_k = u(t_k)$.

To determine Δ_k, we agree to allow only a maximum error of ε in the value of u_{k+1}. Since

(4.75)
$$u_{k+1} = u_k + \Delta_k u'(t_k) + \frac{\Delta_k^2}{2} u''(t_k) + \dots,$$

we see that this requires, to smallest order terms,

(4.76)
$$\frac{\Delta_k^2}{2} |u''(t_k)| \le \varepsilon,$$

or

(4.77)
$$\Delta_k \le \left(\frac{2\varepsilon}{|u''(t_k)|}\right)^{1/2}.$$

Turning to the differential equation, we see that

(4.78)
$$\Delta_k \le \left(\frac{\varepsilon}{|g_u g + g_t|}\right)^{1/2},$$

where g, g_u, and g_t are evaluated at t_k.

In place of this requirement, we may impose the more lenient requirement that the percentage error be less than ε. The resulting inequality is then

(4.79)
$$\Delta_k \le \left(\frac{\varepsilon|u|}{|g_u g + g_t|}\right)^{1/2}.$$

Using the recurrence relation of (4.71), or more complex expressions, we can do even better if needed. In place of using a variable interval, we could use a variable recurrence relation. Here, a more accurate relation would be used for intervals in which the higher derivatives have appreciable values.

4.20 An Example

Consider, as an example of this technique, the equation

(4.80)
$$u' = u^2 + 1, \quad u(0) = 0,$$

whose solution is $u = \tan t$. In the neighborhood of $t = \pi/4$, $u \simeq 1$ and $u' \simeq 2$. The step Δ_k is then bounded by the quantity

(4.81)
$$\left(\frac{\varepsilon}{(u^2 + 1)2u}\right)^{1/2} \simeq \left(\frac{\varepsilon}{4}\right)^{1/2},$$

if we are interested in absolute error, and in this case, by the same quantity if we are concerned with percentage error.

If $\varepsilon = 10^{-4}$, then Δ_k can be taken to be $10^{-2}/2$. At $t = \pi/3$, $u = \tan t = \sqrt{3}$, and $u^2 + 1 = 4$. Then if we are interested in a bound on the percentage error, we have the inequality

$$(4.82) \qquad \Delta_k \leq \left(\frac{\varepsilon}{2(u^2 + 1)}\right)^{1/2} = \left(\frac{\varepsilon}{8}\right)^{1/2}.$$

Very much in the same vein, let us note that the scalar differential equation

$$(4.83) \qquad \frac{du}{dt} = g(u)h(u), \quad u(0) = c,$$

can be approximated to by the difference relation

$$(4.84) \qquad u\left(t + \frac{\Delta}{g(u)}\right) = u(t) + h(u)\Delta.$$

The term $\Delta/g(u)$ automatically introduces a variable grid dependent upon the nature of the solution. Similar representations can be obtained for multidimensional systems.

Writing the right-hand side of (4.83) in the form $g(u)h(u)$ adds a degree of flexibility to the algorithm.

A further application of functional equations as opposed to the usual technique of difference equations in the computational solution of partial differential equations will be found in 5.19.

Bibliography and Comments

4.1 We follow Chapter 8 of the book *Dynamic Programming*, previously cited. For an application of these techniques to a variational problem involving a function of two variables, see

R. Bellman and H. Osborn, "Dynamic programming and the variation of Green's functions," *J. Math. and Mech.*, vol. 7, 1958, pp. 81–86.

4.3 It turns out that this equivalence may also be viewed as an equivalence between Fermat's approach to physical processes and Huygen's approach. For a preliminary discussion, see

A. J. De Witte, "Equivalence of Huygen's approach and Fermat's principle in ray geometry," *Amer. J. Phys.*, vol. 27, 1959, pp. 293–301.

A more extensive discussion may be found in the volume by E. T. Whittaker referred to at the end of Chapter Two.

4.4 For a rigorous discussion of the derivation of equations (4.7) and (4.10), see

H. Osborn, "On the foundations of dynamic programming," *J. Math. and Mech.*, vol. 8, 1959, pp. 867–872.

4.6 For a discussion of these matters, see the results of H. Osborn given in Chapter 7 of

R. Bellman, *Dynamic Programming of Continuous Processes*, The RAND Corporation, Report R-271, August, 1954.

4.8–4.9 The results of these sections are discussed in detail in

S. Dreyfus, *Dynamic Programming and the Calculus of Variations*, Journal of Mathematical Analysis and Applications, 1960.

Dreyfus obtains the classical Hamilton-Jacobi equations from the principle of optimality in a quick and simple way.

Equating the two forms of the solutions of the variational problem yields many interesting identities. See the references to work by Bellman, Lehman, and Osborn in the Bibliography for 5.21.

4.10 Rigorous treatment of the limiting behavior of the discrete process as $\Delta \to 0$ will be found in

R. Bellman, *Dynamic Programming*, Princeton Univ. Press, Princeton, N.J., 1957, Chapter IX,

where results due to W. Fleming are presented; and in

R. Bellman, "Functional equations in the theory of dynamic programming —VI: a direct convergence proof," *Ann. Math.*, vol. 65, 1957, pp. 215–223.

H. Osborn, "The problem of continuous programs," *Pacific J. Math.*, vol. 6, 1956, pp. 721–731,

where methods entirely independent of the calculus of variations are used. This is important, since it means that similar techniques can be used to treat the limiting behavior of discrete stochastic and adaptive control processes, as well as that of multistage games. These are multistage decision processes for which no classical theory exists.

We are so accustomed to treating continuous processes as limits of discrete processes that we may tend to forget that this was once a startlingly original idea, put forth as a hypothesis. It is called Boscovich's Hypothesis. See

R. J. Boscovich, *Treatise on Natural Philosophy*, Venice, 1758, and the article on "Mechanics" in the *Encyclopedia Brittannica*.

4.12 We follow the presentation in

R. Bellman, "Notes on control processes—I: on the minimum of maximum deviation," *Q. Appl. Math.*, vol. 14, 1957, pp. 419–423.

For an application of these ideas, see

M. Ash, R. Bellman, and R. Kalaba, "On control of reactor shutdown involving minimal xenon poisoning," *Nuclear Sci. and Eng.*, vol. 6, 1959, pp. 152–156.

The fact that the linear function $c_1 + c_2 + \ldots + c_N$ and the maximum function $M(c) = \text{Max}\,[c_1, c_2, \ldots, c_N]$ satisfy the same functional equation at first sight appears strange. Consider, however, the function $T_r(c) = (c_1^r + c_2^r + \ldots + c_N^r)^{1/r}$, for $c_i \geq 0$, $r > 0$. It is easy to see that

$$T_r[T_r(c_1, c_2, \ldots, c_{N-1}), c_N] = T_r(c_1, c_2, \ldots, c_N).$$

For $r = 1$, we have the linear transformation, while

$$M(c) = \lim_{r \to \infty} T_r(c).$$

The fact that the maximum value can be taken as the limit of more tractable analytic operations is frequently used to obtain properties of the maximum in a simple fashion. For an application to the study of the heat equation, see

R. Bellman, "A property of summation kernels," *Duke Math. J.*, vol. 15, 1948, pp. 1013–1018.

S. Bochner, "Quasi-analytic functions, Laplace operators and positive kernels," *Ann. Math.*, vol. 51, 1950, pp. 68–91.

R. Bellman and E. F. Beckenbach, *Inequalities*, Ergebnisse der Mathematik, 1960.

It can also be used to obtain the three-circles theorem of Hadamard, the Riesz-Thorin convexity theorem, and so on.

4.15 For a discussion of this problem, see

R. Bellman, I. Glicksberg, and O. Gross, "On the 'bang-bang' control problem," *Q. Appl. Math.*, vol. 14, 1956, pp. 11–18.

J. P. LaSalle, "On time optimal control systems," *Proc. Nat. Acad. Sci. USA*, vol. 45, 1959, pp. 573–577.

L. S. Pontrjagin, "Some mathematical problems arising in connection with the theory of optimal automatic control systems," *Proc. Conf. on Basic Problems in Automatic Control and Regulation*, Academy of Sciences, Moscow, 1957.

A. M. Hopkin, "A phase-plane approach to the compensation of saturating servomechanisms," *AIEE Trans.*, vol. 70, part 1, 1951, pp. 631–639.

A. Feldbaum, "Optimal processes in automatic control systems," *Automat. i. Telemeh.*, vol. 14, no. 6, 1953, pp. 712–728.

R. Kalman and J. Bertram, "General synthesis procedure for computer control of single and multiloop linear systems," *Proc. Computers in Control Systems Conf.*, AIEE Special Publ. T-101, 1958.

N. Krasovskii, "On the theory of optimal regulation," *Automat. i. Telemeh.*, vol. 18, no. 11, 1957, pp. 960–970.

N. Krasovski, "On a problem of optimal regulation," *Prik. Mat. i. Meh.*, vol. 21, no. 5, 1957, pp. 670–677.

V. Boltyanskii, "The 'maximum' principle in the theory of optimal processes," *Dokl. Akad. Nauk. USSR*, vol. 119, no. 6, 1958, pp. 1070–1073.

L. A. Zadeh, *Computer-Control Systems Technology*, Univ. of California Engineering Extension Lectures, May 18–22, 1959.

C. W. Merriam, III, "A class of optimum control systems," *J. Franklin Institute*, 1959, pp. 267–282.

C. W. Merriam, III, "Use of a mathematical criterion in the design of adaptive control systems," *Applications and Industry*, No. 46, 1960, pp. 506–519.

For a detailed analysis of the Pontrjagin maximum principle and a proof of its equivalence to the functional equation yielded by the principle of optimality, see

L. T. Rozonoer, "L. S. Pontrjagin's maximum function principle in its application to the theory of optimum systems—I," *Automatika i Telemakhanika*, vol. 20, 1959, pp. 1320–1334,

"—II," vol. 20, 1959, pp. 1441–1458,

"—III," vol. 20, 1959, pp. 1561–1578.

The first two in the series present a proof of the Pontrjagin principle based upon the classical calculus of variations, while the third examines the relation between Pontrjagin's methods and the results given above. Pontrjagin's results were developed in 1956. Many references will be found in Part I.

4.16 For applications of dynamic programming:
to trajectories, see

T. Cartaino and S. Dreyfus, "Application of dynamic programming to the airplane minimum time-to-climb problem," *Aero. Eng. Review*, 1957; See also, J. S. Fiorentino, "Dynamic Programming and the Solution of Missile Control Problems, May, 1960, General Dynamics, ERR–PO–013.

to satellite launching, see

R. Bellman and S. Dreyfus, "An application of dynamic programming to the determination of optimal satellite trajectories," *J. British Interplanetary Soc.*, vol. 17, 1959–60, pp. 78–83.

This treats a problem first posed in

D. E. Okhotsimskii and T. M. Eneev, *J. British Interplanetary Soc.*, Jan.–Feb., 1958;

to paths of minimum length through networks, see

R. Bellman, "On a routing problem," *Q. Appl. Math.*, vol. 16, 1958, pp. 87–90;

R. Kalaba, "On some communication network problems," *Proc. Sym. Appl. Math.*, April, 1958;

to k-th shortest paths, see

R. Bellman and R. Kalaba, "On k-th best policies," *J. Soc. Ind. Appl. Math.*, to appear;

to "soft" landings, see

R. Bellman and J. M. Richardson, "On the application of dynamic programming to a class of implicit variational problems," *Q. Appl. Math.*, vol. 17, 1959, pp. 231–236.

to chemical process control, see

R. Aris, R. Bellman, and R. Kalaba, "Some applications of dynamic programming to chemical processes," to appear;

to reactor control, see

M. Ash, R. Bellman, and R. Kalaba, "On control of reactor shutdown involving minimal xenon poisoning," *Nuclear Sci. and Eng.*, vol. 6, 1959, pp. 152–156.

For a dynamic programming treatment of the optimum staging of rockets, see

R. P. Ten Dyke, "Computation of rocket step weights to minimize initial gross weight," *Jet Propulsion*, vol. 28, 1958, pp. 338–340.

For the conventional approach which encounters certain computational difficulties, see

C. H. Builder, "General solution for optimization of staging of multistaged boost vehicles," *ARS J.* (formerly *Jet Propulsion*), vol. 29, 1959, pp. 497–499,

where references to a number of earlier papers may be found.

An excellent expository paper is

D. F. Landen, "Mathematical problems of astronautics," *Math. Gazette*, vol. 41, 1957, pp. 168–179.

4.17 An excellent discussion of the brachistochrone problem is contained in Chapter III of

G. A. Bliss, *Lectures on the Calculus of Variations*, Univ. of Chicago Press, 1946.

The problem was first posed by Galileo and solved by the Bernoulli brothers.

It was shown by Caratheodory that extremals of general variational problems can be considered to be "curves of quickest descent." This very important idea would seem to be of use in connection with the approximate solution of variational problems. However, it has not as yet been exploited.

C. Caratheodory, *Uber diskontinuirliche Losungen in der Variationsrechnung*, Diss. Gottingen, 1904.

——————, *Variationsrechnung und partielle Differentialgleichungen erster Ordnung*, 1935, pp. 249–251.

A large number of physical processes can be cast in the format of multistage processes, not only over time, the classical approach, but with respect to other physical parameters such as length, energy, and so on.

For applications of this concept to radiative transfer, see

S. Chandrasekhar, *Radiative Transfer*, Oxford, 1950.

R. Bellman and R. Kalaba, "On the principle of invariant imbedding and propagation through inhomogeneous media," *Proc. Nat. Acad. Sci. USA*, vol. 42, 1956, pp. 629–632;

to neutron transport theory,

R. Bellman, R. Kalaba, and G. M. Wing, "On the principle of invariant imbedding and one-dimensional neutron multiplication," *Proc. Nat. Acad. Sci. USA*, vol. 43, 1957, pp. 517–520;

to random walk and scattering,

R. Bellman and R. Kalaba, "Random walk, scattering, and invariant imbedding—I: one-dimensional case," *Proc. Nat. Acad. Sci. USA*, vol. 43, 1957, pp. 930–933;

to wave propagation,

R. Bellman and R. Kalaba, "Invariant imbedding, wave propagation and the WKB approximation," *Proc. Nat. Acad. Sci. USA*, vol. 44, 1958, pp, 317–319.

For a survey of the preceding results, see

R. Bellman, R. Kalaba, and G. M. Wing, "Invariant imbedding and mathematical physics—I: particle processes," *J. Math. Physics*, July, 1960.

4.18 For numerical solution of differential equations, see

L. Collatz, *Numerische Behandelung von Differentialgleichungen*, Z. Aufl. Berlin, 1955.

This application of feedback control to choose the size of the time-step has been used in the computing field for some time, as was kindly pointed out to me by M. Prestrud. We have not been able to ascertain who first applied this technique.

An interesting paper unifying classical and new variational principles in mathematical physics is

S. Altschuler, "Variational principles for the wave function in scattering theory," *Phys. Rev.* (2), vol. 109, 1958, pp. 1830–1836.

For a further discussion of the Hamilton-Jacobi formalism, see

N. Cetaev, On the extension of the optical-mechanical analogy, J. Appl. Mech., 22 (1958) pp. 678–681.

For additional papers on the connection between Huygens' and Fermat's principles, referred to in the discussion of 4.3, see the references given in the review of de Witte's paper by N. Chako in Mathematical Reviews, Vol. 21 (1960), p. 870, No. 4736.

CHAPTER V

COMPUTATIONAL ASPECTS OF DYNAMIC PROGRAMMING

The Kingdom of Number is all boundaries
Which may be beautiful and must be true
To ask if it is big or small proclaims one
The sort of lover who should stick to faces.

W. H. Auden
"Numbers and Faces"

5.1 Introduction

We have seen in the foregoing chapter that the variational problem of minimizing the integral

$$(5.1) \qquad J(v) = \int_0^T g(u,\, v)\, dt$$

subject to constraints of the form

$$(5.2) \qquad \frac{du}{dt} = h(u,\, v), \quad u(0) = c,$$

can be transformed into the problem of determining a sequence of functions $\{f_N(c)\}$ by means of a recurrence relation of the form

$$(5.3) \qquad f_1(c) = \operatorname*{Max}_{v} [g(c,\, v)\Delta],$$

$$f_N(c) = \operatorname*{Max}_{v} [g(c,\, v)\Delta + f_{N-1}[c + h(c,\, v)\Delta]],$$

$N = 2, 3, \ldots.$

In this chapter we wish to discuss the computational feasibility of a solution based on an algorithm of this type, how this method overcomes some of the difficulties encountered by classical methods we have stressed in Chapter I, and what new difficulties arise.

As we shall see, although these techniques enable us to make some progress, there is still much to be done and many new ideas to be sought before we can rest content with our labors.

At the end of the chapter we briefly indicate the applicability of some of our computational techniques to other classes of partial differential equations. These applications are suggested by the dynamic programming approach.

5.2 The Computational Process—I

Let us now describe in detail how the sequence $\{f_N(c)\}$ can be computed using the relations in (5.3). We are now thinking purely along numerical lines in terms of operations involving the use of digital computers. In subsequent chapters we shall discuss various analytic devices that can be used to increase the efficiency of functional equation methods.

In order to classify a particular member of the sequence, say $f_k(c)$, as known, we wish to specify the values taken by this function for all values of c. Since we clearly cannot tabulate *all* values, what we do is decide to evaluate $f_k(c)$ at a certain subset of c-values, and then use these values in some fashion to represent the function for all values of c.

Generally, these c-values have the form

$$(5.4) \qquad c = M\,\delta,\,(M+1)\,\delta,\,\ldots,\,(N-1)\,\delta,\,N\,\delta,$$

where $[M\,\delta,\,N\,\delta]$ is the interval of c-values of greatest interest. Given the functional values at the points $c = j\delta$, $j = M,\,M+1,\,\ldots,\,N$, other values of $f_k(c)$ in the interval $[M\,\delta,\,N\,\delta]$ are determined by using the value of $f_k(c)$ at the nearest point of the form $j\,\delta$, or by using interpolation formulas of various degrees of sophistication. Furthermore, values outside of this interval can be determined by means of extrapolation.

Thus, we may set

$$(5.5) \qquad f_k(c) = f_k(\delta[c/\delta]),\quad \delta[c/\delta] \le c \le \delta([c/\delta]+1),$$

where $[x]$ denotes the greatest integer less than or equal to x. Or, if $a = \delta[c/\delta]$, $b = \delta([c/\delta]+1)$, we may use a linear interpolation formula of the type

$$(5.6) \qquad f_k(c) = f_k(a) + (c-a)\frac{[f_k(b)-f(a)]}{(b-a)},\quad a \le c \le b,$$

or, finally, four-point, six-point, or more accurate interpolation. The type of interpolation used depends upon the accuracy desired, the time required, and the memory of the computing device that is available.

Given a c-interval of interest, $[a, b]$, the choice of δ determines the number of different points at which we must tabulate the functional value of $f_k(c)$. On one hand, for purposes of accuracy, we would like to take δ small; on the other hand, from the standpoint of computing time, and availability of computer memory, we would like δ to be large. At the present time, there is very little theory behind the choice of δ. It is mainly a matter of experience, and trial and error, both of which are occasionally painful.

Having decided upon the range of c-values, we must decide upon the possible values of v which we shall admit, since it is clearly impossible by numerical methods alone to maximize over *all* values of v. We have previously discussed the limited applicability of the methods of calculus in determining maximum values. Consequently, in applying the formulas in (5.3), we wish to determine the maximum not by means of derivatives, but rather by a straightforward *search technique*. By this we mean that we wish to evaluate the functions of v appearing in the right-hand side of the equations (5.3) for a sufficiently large set of v-values and then by direct comparison of values—a simple process for digital computers—determine the absolute maximum.

Sometimes, v is constrained to be a member of a discrete set of values, $\{v_1, v_2, \ldots, v_R\}$, and sometimes it is constrained to an interval $[a_1, b_1]$. Let us consider the case of continuous variation, since that of discrete variation is simpler. To treat the situation where $a_1 \leq v \leq b_1$, we allow v to assume only the values

(5.7) $\qquad a_1 = M_1\,\delta_1, (M_1 + 1)\,\delta_1, \ldots, (N_1 - 1)\,\delta_1, N_1\,\delta_1 = b_1.$

Once again, we would like to choose δ_1 to be small for the sake of accuracy, but large from the standpoint of saving time. Experience and trial and error determine satisfactory choices, since there is little or no theory to guide one.

In a later section, 5.17, devoted to the subject of sequential search, itself a dynamic programming process, we shall discuss how certain analytic results can be utilized to improve upon this most direct search process.

5.3 The Computational Process—II

Having chosen the grid of c-values and the set of allowable v-values, we turn to (5.3) and compute the function $f_1(c)$. To do this, we take the first c-value, say $M\,\delta$, and evaluate

(5.8) $\qquad f_1(M\,\delta) = \underset{v}{\operatorname{Max}}\,[g(M\,\delta, v)\Delta],$

$\qquad\qquad\qquad = \underset{M_1 \leq i \leq N_1}{\operatorname{Max}}\,[g(M\,\delta, i\,\delta_1)\Delta].$

We accomplish this by evaluating first the quantity $g(M\,\delta, M_1\,\delta_1)$ and then $g(M\,\delta, (M_1 + 1)\,\delta_1)$, either by means of analytic representations or from a table of values stored in the memory of the computer. These two values are then compared and the larger value kept. Then $g[M\,\delta, (M_1 + 2)\,\delta_1]$ is evaluated and compared with the largest value previously retained. This process is continued for the allowable set of v-values. The largest value obtained is clearly an *absolute maximum*.

In obtaining this maximum, we also obtain the value of v which maximizes. It is possible that several values of v yield the absolute maximum. In this case, we can have the machine retain all maximizing values, or else keep the first one.

Having run through this process for the value $c = M\,\delta$, we then perform exactly the same sequence of operations for $c = (M + 1)\,\delta$, $c = (M + 2)\,\delta$, and so on.

In this way, we not only obtain the sequence of values $\{f_1(k\,\delta)\}$, $k = M, M + 1, \ldots, N$, but also the set of v-values which yield the maximum in (1.3). This set of v-values determines a function which we call $v_1(c)$, determined by the sequence of values $\{v_1(k\,\delta)\}$, $k = M, M + 1, \ldots, N$. Values of $v_1(c)$ not given explicitly may again be determined by interpolation. Let us call this function, as before, the *policy-function*. As noted above, it need not be single-valued.

5.4 The Computational Process—III

Once $f_1(c)$ has been determined, we are ready to determine the function $f_2(c)$ by way of (5.3). This yields the recurrence relation

$$(5.9) \qquad f_2(c) = \operatorname*{Max}_{v} \left[g(c, v)\Delta + f_1[c + h(c, v)\Delta] \right].$$

Since $f_1(c)$ can now be considered to be a known function, we repeat the previous operations and evaluate $f_2(c)$ and $v_2(c)$ in the fashion described above. There is, however, a possible hitch!

5.5 Expanding Grid

We agreed to determine the function $f_1(c)$ by evaluating it at the points $i\,\delta$, $M \leq i \leq N$. Turning to (5.9), we see that the evaluation of $f_2(N\,\delta)$ involves the value of $f_1(c)$ at the point $c = N\,\delta + h(N\,\delta, v)\Delta$. If $h(c, v)$ is uniformly positive, as is often the case, we see that the evaluation of $f_2(c)$ at $c = N\,\delta$ requires the knowledge of $f_1(c)$ at a point c greater than $N\,\delta$, which is to say outside the original c-grid. This phenomenon is sometimes called the *menace of the expanding grid*.

In order to obtain $f_2(c)$ at the set of values $i\,\delta$, $M \leq i \leq N$, we must then determine $f_1(c)$ at a larger set of values.

5.6 The Computational Process—IV

Once the function $f_2(c)$ and its associated policy function $v_2(c)$ have been determined, we repeat this process to determine $f_3(c)$ and $v_3(c)$, $f_4(c)$ and $v_4(c)$ and so on.

As pointed out above, it is quite possible that in order to determine $f_N(c)$ on a specified interval $[a_N, b_N]$, we may have to determine $f_{N-1}(c)$ on a larger interval $[a_{N-1}, b_{N-1}]$, and so on, until finally $f_1(c)$ is determined on a very much larger interval, $[a_1, b_1]$. This is a familiar situation in the theory of partial differential equations. Although we always try very hard to avoid it, sometimes nothing can be done.

Without going into details of coding and programming, it is worth pointing out that the general type of computational process outlined above is ideally suited for digital computers since it involves performing the same type of operation over and over.

We have accomplished our aim of reducing the numerical solution of a variational process to an iterative procedure.

Associated with this procedure, again as in the case of ordinary and partial differential equations, are a number of intriguing and significant convergence and stability questions. It is clear that we are *not* generating the original sequence, $\{f_k(c)\}$, but rather a new sequence, $\{\phi_k(c)\}$. This new sequence differs from $\{f_k(c)\}$ in two significant ways. In the first place, it agrees with the old sequence only at a finite set of values, at most, and it is obtained not by means of an actual maximization, but by means of a maximization over a finite set of values. It is essential that we estimate the error involved in these two approximations. This requires a stability analysis. If this is carried out, and all goes well, it should yield, as a by-product, the result that as $\delta \to 0$, $\delta_1 \to 0$, we have $\phi_k(c) \to f_k(c)$.

We shall not discuss any of these matters here.

5.7 Obtaining the Solution from the Numerical Results

As a result of these computations, we obtain a set of functions $\{f_k(c)\}$, $\{v_k(c)\}$, evaluated at the points $M\,\delta$, $(M+1)\,\delta$, ... , $N\,\delta$. We thus obtain a chart of values of the following form:

c	$f_1(c)$	$v_1(c)$	$f_2(c)$	$v_2(c)$	$\cdots$	$f_N(c)$	$v_N(c)$
$M\,\delta$	---	---	---	---	---	---	---
$(M+1)\,\delta$	---	---	---	---	---	---	---
$(N-1)\,\delta$	---	---	---	---	---	---	---
$N\,\delta$	---	---	---	---	---	---	---

How do we use this array to solve a particular variational problem in which c and N have given numerical values?

We start an N-stage process in state c. Referring to the tables above, we see that the optimal choice of v is given by the value $v_N(c)$. Consequently, the new c-value will be given by

$$(5.10) \qquad c_1 = c + h[c, v_N(c)]\Delta.$$

Having made the initial choice, we face an $(N-1)$-stage process in state c_1. Referring to the tables, we see that the optimal choice of v at the second stage is given by $v_{N-1}(c_1)$. The new state is then given by

$$(5.11) \qquad c_2 = c_1 + h[c_1, v_{N-1}(c_1)]\Delta.$$

Continuing in this way, we obtain the succession of states, $\{c_i\}$, and the succession of v-values. We thus obtain the solution to any specific problem. This can be done by hand, using the tables given above, or it can be done by machine, using the foregoing recurrence relations.

5.8 Why is Dynamic Programming better than Straightforward Enumeration?

At this point, the reader may ask, and should ask, what advantage has the complicated search method described above over a straightforward enumerative technique which explores *all* possible policies?

To answer this question briefly, consider a particular case where v can assume a hundred different values at each stage, and a variational problem involving a 20-stage process. The total number of possible policies is then $(100)^{20} = 10^{40}$. To give some idea of the magnitude of this number, suppose that the evaluation of a particular policy takes a microsecond—a charitable estimate. Then we require 10^{34} seconds, $10^{34}/60$ minutes, $10^{34}/3600$ hours, $10^{34}/(3600)(24)$ days, or finally, $10^{34}/(3600)(24)(365)$ years. Since

$$(5.12) \qquad \frac{10^{34}}{(3600)(24)(365)} \geq \frac{10^{34}}{(10000)(100)(1000)} = 10^{25},$$

we see that this straightforward enumeration of policies would require at least 10^{25} years. Since the age of the earth is of the order of 5 billion years $= 5 \times 10^9$ years, we see that we would require a time considerably in excess of the age of the earth.

So much for the idea of direct enumeration of possibilities! Yet this problem is a trivial one from the standpoint of dynamic programming.

This then furnishes one answer to the question. Clearly in solving variational problems by means of dynamic programming we are doing something a bit better than direct enumeration. But what?

The answer is the following. When we use (5.3) and test some particular value of v, we do not allow *all* possible continuations, but only those which correspond to optimal continuations from the new state, $c_1 = c + h(c, v)\Delta$. This is what the principle of optimality does for us. It automatically eliminates myriad policies of no importance.

5.9 Advantages of Dynamic Programming Approach

In the following sections we wish to show how the functional equation technique of dynamic programming overcomes the formidable difficulties which balk the application of the classical techniques to so many significant problems. We will discuss in turn the questions of absolute maximization, two-point boundary-value problems, constraints, non-analyticity, and implicit variational problems.

90

Having disposed of these, we shall then bare the skeleton in the closet which prevents us from regarding all problems in the calculus of variations, and thus in feedback control, optimal trajectories, and so on, as routinely solved. We are, however, closing in on the more realistic and complex problems from several directions, analytically and technologically. These more sophisticated matters will be discussed in subsequent chapters.

Nonetheless, just as we cannot expect a final problem, so we cannot expect a final solution. Each solution is an impetus to a more complex problem, which in turn calls forth a greater effort, et cetera, et cetera, et cetera.*

5.10 Absolute Maximum versus Relative Maximum

It is clear from the discussion we have given of the computational procedure that it is the *absolute maximum* we determine. A simple inductive argument establishes this. Thus, we automatically bypass the difficulties posed by the possible occurrence of relative maxima and stationary points of even more complicated nature.

5.11 Initial Value versus Two-point Boundary Value Problems

The computational technique described above is an iterative one, starting with a known function $f_1(c)$ and determining in turn $f_2(c), \ldots,$ $f_N(c)$. We thus bypass the two-point boundary-value problem of the classical approach. In Chapter VII we shall discuss these matters in detail.

5.12 Constraints

As we have pointed out in Chapter I, the addition of a constraint of the form

$$(5.13) \qquad |y(t)| \leq m, \quad 0 \leq t \leq T,$$

adds an amazing amount of complication to the solution of a problem in the calculus of variations. Let us see what happens in the dynamic programming approach.

The policy function $v = v(c, T)$ is now restricted by the condition

$$(5.14) \qquad |v| \leq m.$$

It follows, as has already been pointed out in 4.13 that the recurrence relations of (5.3) take the form

$$(5.15) \qquad f_1(c) = \underset{|v| \leq m}{\text{Max}} [g(c, v)\Delta],$$

$$f_N(c) = \underset{|v| \leq m}{\text{Max}} [g(c, v)\Delta + f_{N-1}(c + h(c, v)\Delta)].$$

We have seen that constraints of this type are a nuisance as far as calculus is concerned. But what happens to the computational solution we have

---

* To quote from "Anna and the King of Siam."

outlined? *The presence of the constraint simplifies the determination of the solution!*

The reason for this is that by means of the constraint $|v| \leq m$ we have cut down on the allowable choice of policies at each stage. The search process is thus easier to carry out. It is obvious, a priori, that the more restrictions we impose, the smaller is the set of feasible policies. It was the method which introduced the previously cited complications, not the problem.

5.13 Non-analyticity

Finally, let us point out that we have made no use of differential properties of the functions $g(c, v)$ and $h(c, v)$. We have assumed continuity in using the notation $\underset{v}{\text{Max}}$. However, if we restrict c and v to a discrete set of values, not even this condition is required. All that is required is that we be able to determine the values of g and h, given the values of c and v. The technique of dynamic programming is thus ideally suited to treat realistic control processes in which $g(c, v)$ and $h(c, v)$ may exist graphically, determined by experiment, rather than analytically, determined by theory.

When using these methods, there is no need to use linear or quadratic functions and no need to use approximations designed to furnish simple equations rather than designed to solve realistic problems. Not only can we use more accurate descriptions of the physical process in writing the governing equations, but we can also employ more sensitive and accurate criteria. We gave one example of this in the treatment of the minimization of maximum deviation in Chapter IV.

Furthermore, we are no longer required to vary in any continuous fashion. It follows that we can consider far more realistic control processes in which only a finite number of alternatives are available at each stage.

5.14 Implicit Variational Problems

In the discussion of the "bang-bang" control problem in Chapter IV, we indicated the applicability of the functional equation technique to the problem of determining the function $v(t)$, (subject to the constraint

$$(5.16) \qquad |v| \leq m, \quad t \geq 0)$$

which forces the solution of

$$(5.17) \qquad u'' = g(u, u', v), \quad u(0) = c_1, \quad u'(0) = c_2,$$

from its initial state, (c_1, c_2), to $(0, 0)$ in minimum time.

Using the discrete version,

$$(5.18) \quad u_1(n + 1) = u_1(n) + u_2(n)\Delta, \quad u_1(0) = c_1,$$

$$u_2(n + 1) = u_2(n) + g[u_1(n), u_2(n), v(n)]\Delta, \quad u_2(0) = c_2,$$

we showed that the function of minimum time satisfied the equation

$$(5.19) \qquad f(c_1, c_2) = \underset{|v| \leq m}{\text{Min}} f(c_1 + c_2\Delta, \ c_2 + g(c_1, c_2, v)\Delta).$$

Since this is no longer an initial value problem, due to the fact that the unknown function appears on both sides of the equation, it is not obvious as to how to proceed to calculate $f(c_1, c_2)$. We shall discuss two approaches, one in this section and one in the following section, both of interest in many other connections.

To motivate the first approach, let us change our problem, and ask for the policy which minimizes the distance of the solution from the equilibrium position at the end of N stages. Call this minimum distance $\phi_N(c_1, c_2)$. Let us actually use the square of the Euclidean distance to avoid square roots. Setting $\phi_0(c_1, c_2) = c_1^2 + c_2^2$, we have the recurrence relation

$$(5.20) \qquad \phi_N(c_1, c_2) = \underset{|v| \leq m}{\text{Min}} \ \phi_{N-1}(c_1 + c_2\Delta, \ c_2 + g(c_1, c_2, v)\Delta),$$

for $N \geq 1$.

To determine the function $f(c_1, c_2)$, we compute the sequence $\phi_N(c_1, c_2)$ until we find the first value of N for which $\phi_N(c_1, c_2) = 0$, or, for practical purposes, $|\phi_N(c_1, c_2)| \leq \varepsilon$.

It is clear that this function $N(c_1, c_2)$ is precisely the minimum time required to transform the system from state (c_1, c_2) to the state $(0, 0)$.

5.15 Approximation in Policy Space

Let us now consider the following approach to the solution of (5.19). Let $v_0 = v_0(c_1, c_2)$ be an initial approximation to the optimal policy, and let $f_0(c_1, c_2)$ be computed by means of the equation

$$(5.21) \qquad f_0(c_1, c_2) = 1 + f_0[c_1 + c_2\Delta, \ c_2 + g(c_1, c_2, v_0)\Delta].$$

In other words, $f_0(c_1, c_2)$ is the time required to force the system to equilibrium using the approximate policy $v_0(c_1, c_2)$. For example, we may choose v_0 so that the distance from the origin is uniformly decreased.

Let $f_1(c_1, c_2)$ be determined by the relation

$$(5.22) \qquad f_1(c_1, c_2) = \underset{|v| \leq m}{\text{Min}} f_0[c_1 + c_2\Delta, \ c_2 + g(c_1, c_2, v)\Delta],$$

and then, inductively, set

$$(5.23) \qquad f_N(c_1, c_2) = \underset{|v| \leq m}{\text{Min}} f_{N-1}[c_1 + c_2\Delta, \ c_2 + g(c_1, c_2, v)\Delta].$$

It is clear that

$$(5.24) \qquad f_0(c_1, c_2) \geq f_1(c_1, c_2) \geq \ldots \geq f_N(c_1, c_2) \geq \ldots,$$

so that we have monotone approximation, and actual convergence. Since we started by approximating to the policy rather than to the return function, we call this *approximation in policy space*. Actually, it is only one particular application of this method, and a hybrid version at that.

We shall discuss this powerful technique in greater detail in 18.4, where a truer version of approximation in policy space is presented. It is briefly touched upon in Chapter VIII, where a problem of the same abstract form but quite different origin is considered by the same method.

5.16 The Curse of Dimensionality

In view of all that we have said in the foregoing sections, the many obstacles we appear to have surmounted, what casts the pall over our victory celebration? It is the curse of dimensionality, a malediction that has plagued the scientist from earliest days.

Suppose that we decide to treat the general vector variational problem by these techniques. Let

$$(5.25) \qquad f(c, T) = \operatorname*{Min}_{y} \int_0^T g(x, y) \, dt,$$

where

$$(5.26) \qquad \frac{dx}{dt} = h(x, y), \quad x(0) = c,$$

x is N-dimensional and y is M-dimensional.

Then a discrete version, along the foregoing lines, yields the relation

$$(5.27) \qquad f_N(c) = \operatorname*{Min}_{y} [g(c, y)\Delta + f_{N-1}(c + h(c, y)\Delta)].$$

Abstractly we have the same algorithm regardless of the dimension of x. But what happens to the computational feasibility?

A function of one variable, $u = u(t)$, can be readily visualized as a curve in the plane, and readily tabulated. A function of two variables, $u = u(s, t)$, a surface in three dimensions, is a great deal more troublesome both to visualize and tabulate, and functions of three or more variables are impossible to visualize geometrically and exceedingly difficult to tabulate.

It is more than a matter of inconvenience. If $u = u(t)$ is specified over $0 \leq t \leq 1$ by its values at the points $k(.01)$, $k = 0, 1, 2, \ldots, 99$, we must tabulate 100 values. If $u = u(s, t)$ is specified over $0 \leq s, t \leq 1$ by a grid of similar type in s and t, we now require $100 \times 100 = 10^4$ values. Similarly, a function of three values defined over the same grid requires 10^6 values.

Since current machines have fast memories capable of storing only 3×10^4 values, and contemplated machines over the next ten years may go up only to 10^6 values, we see that multidimensional variational problems cannot be solved *routinely* because of the memory requirements.

This does not mean that we cannot attack them. It merely means that we must employ some more sophisticated techniques. We shall discuss some of these in later chapters. This challenge adds to the interest of these problems, and points up the remark made previously that we can expect no final solution.

5.17 Sequential Search

One way of cutting down on the time required for the computational solution is to speed up the search process for the maximum described in 5.3. It turns out that in some cases there are techniques we can apply which reduce by a factor of one hundred or more, the time that would be required by straightforward search. Nevertheless, a great deal of work remains to be done in this field, since only the simplest problems have been resolved, and even these have required a great deal of ingenuity.

5.18 Sensitivity Analysis

Let us note finally that the dynamic programming approach, as a by-product, yields the solution as a function of two basic parameters, the initial state and the duration of the process. This means that whenever we solve one problem, we automatically solve a family of problems.

At first sight this would seem to be quite wasteful. Actually, the situation is just the reverse. Considering the many assumptions that go into the construction of mathematical models, the many uncertainties that are always present, we must view with some suspicion any particular prediction.

One way to obtain confidence is to test the consequences of various changes in the basic parameters. This *stability anaylsis* or *sensitivity analysis* is always essential in evaluating the worth of results obtained from a particular model.

Consequently, we never want isolated solutions. We may be required to "make-do" with these, but we always prefer a family of solutions to a family of problems. If, as is true in some cases, "plus change, plus même" describes the situation, then we may feel that we have results of some significance.

5.19 Numerical Solution of Partial Differential Equations

In the preceding chapter, we observed that the problem of minimizing the integral

$$(5.28) \qquad J(y) = \int_0^T g(u, v) \, dt$$

subject to constraints of the form

$$(5.29) \quad (a) \quad \frac{du}{dt} = h(u, v),$$

$$(b) \quad u(0) = c,$$

was, on the one hand, equivalent to that of determining the solution of the nonlinear partial differential equation

$$(5.30) \qquad \frac{\partial f}{\partial T} = \underset{v}{\text{Max}} \left[g(c, v) + h(c, v) \frac{\partial f}{\partial c} \right],$$

$$f(c, 0) = 0,$$

and on the other hand, could be treated computationally by means of the recurrence relation

$$(5.31) \qquad f_{N+1}(c) = \underset{v_0}{\text{Max}} \left[g(c, y_0)\Delta + f_N(c + h(c, y_0)\Delta) \right].$$

It follows that we can use (5.31) to obtain the numerical solution of (5.30) in place of the conventional difference techniques. We wish to discuss some further applications of this idea.

5.20 A Simple Nonlinear Hyperbolic Equation

Consider the equation

$$(5.32) \qquad u_t = g(u)u_x + h(u, x, t),$$

$$u(x, 0) = f(x).$$

One way to obtain the computational solution of this equation is to replace the continuous range of x and t values by a discrete grid, say

$$(5.33) \qquad x = k\,\delta, \quad k = -N, -N + 1, \ldots, N - 1, N,$$

$$t = r\Delta, \quad r = 0, 1, \ldots,$$

and the partial differential equation of (5.32) by a partial difference equation such as

$$(5.34) \qquad \left[\frac{u(x, t + \Delta) - u(x, t - \Delta)}{2\Delta} \right]$$

$$= g(u) \left[\frac{u(x + \delta, t) - u(x - \delta, t)}{2\delta} \right] + h(u, x, t).$$

Observe that simple and direct as this procedure is, it introduces some complications in the form of the ratio Δ/δ. Although both δ and Δ are "small," the ratio Δ/δ may be large.

If the behavior of u as a function of x is erratic, we may be forced to use small steps in x, i.e. small values of δ, and thus small values of Δ. This, however, greatly increases the time required for computation—a quandary!

One way to avoid this difficulty is to follow the pattern set by the recurrence relation of (5.31). To illustrate this, consider the equation

$$(5.35) \qquad u_t = -uu_x, \quad u(x, 0) = f(x).$$

$$f_T = g_1 + g_2 f_c + g_3 f_c^2$$

Consider the recurrence relation

(5.36) $$u(x, t + \Delta) = u(x - u(x, t)\Delta, t),$$

where t assumes the values $0, \Delta, 2\Delta, \ldots$ Considering Δ to be a small quantity, we see that (5.36) is equivalent to the relation

(5.37) $$u(x, t) + \Delta u_t + O(\Delta^2) = u(x, t) - u(x, t)u_x \Delta + O(\Delta^2),$$

or

(5.38) $$u_t = -uu_x + O(\Delta).$$

The computational method presented in (5.36) does not involve the notion of an x-grid, since the values of $x - u(x, t)\Delta$ will not be distributed in any regular fashion. Tabulating the values of $u(x, t)$ at values $x = 0, \delta, 2\delta, \ldots,$ we compute the value of $u[x - u(x, t)\Delta, t]$ by using an interpolation formula.

This method has been applied to the equation used in (5.35), and to some others, and favorable results have been obtained.

5.21 The Equation $f_T = g_1 + g_2 f_c + g_3 f_c^2$

Suppose in (5.30) that

(5.39) $$g(c, v) = g_1(c) - v^2,$$
$$h(c, v) = g_2(c) + vg_3(c).$$

Then, we have the equation

(5.40) $$\frac{\partial f}{\partial T} = \underset{v}{\text{Max}} \left[-v^2 + vg_3(c) \frac{\partial f}{\partial c} \right] + g_1(c) + g_2(c) \frac{\partial f}{\partial c}.$$

Since the maximum occurs at

(5.41) $$v = \tfrac{1}{2} g_3(c) \frac{\partial f}{\partial c},$$

and its value is

(5.42) $$g_3^2(c) \left(\frac{\partial f}{\partial c} \right)^2 \Big/ 4,$$

we obtain the quadratically nonlinear partial differential equation

(5.43) $$\frac{\partial f}{\partial T} = g_1(c) + g_2(c) \frac{\partial f}{\partial c} + \frac{g_3^2(c)}{4} \left(\frac{\partial f}{\partial c} \right)^2.$$

In place of solving this equation by means of differences, we can use the recurrence relation derived from (5.40),

(5.44) $$f_{N+1}(c) = \underset{y_0}{\text{Max}} \left[(g_1(c) + y_0^2)\Delta + f_N(c + g_2(c)\Delta + y_0 g_3(c)\Delta) \right],$$

97

which may be simpler and more accurate. Furthermore, we possess the advantage of knowing approximately where the minimum value is attained.

We can, of course, obtain similar remarks for the equation derived from the case where $g(c, v) = g_1(c) + v^{2k}$. These ideas are intimately related to the technique of quasi-linearization discussed in Chapter XII.

Bibliography and Comments

5.1 The numerical solution of the functional equations of dynamic programming is considered in detail in

R. Bellman and S. Dreyfus, *Computational Aspects of Dynamic Programming*, Princeton Univ. Press, Princeton, N.J., 1961.

An excellent elementary discussion of the operation of digital computers may be found in

J. von Neumann, *The Computer and the Brain*, Yale Univ. Press, 1956.

The second part of this short monograph is extremely valuable in connection with our later work in adaptive processes.

5.6 A rigorous treatment of the problem of the convergence of the discrete version to the continuous version may be found in

R. Bellman, "Functional equations in the theory of dynamic programming," *Ann. Math.*, vol. 65, 1957, pp. 215–223.

H. Osborn, "The problem of continuous programs," *Pacific J. Math.*, vol. 6, 1956, pp. 721–731.

5.7 A very ingenious way of obtaining the solution to a particular problem has been devised by S. Dreyfus. It is presented in the first book cited above.

5.8 What the principle of optimality asserts is that there is nothing lost in restricting our attention to the subclass of continuations among which we know a priori the optimal policy must lie. For a discussion of this important idea in the context of the calculus of variations, see

R. Courant and D. Hilbert, *Methoden der Mathematische Physik*, vol. 1, pp. 231–233, Interscience.

5.10 It is interesting to note that the functional equation technique can be used to obtain second best, third best, and so on, policies; see

R. Bellman and R. Kalaba, "On k-th best policies," *J. Soc. Ind. Appl. Math.*, to appear.

5.17 See the work by S. Johnson, O. Gross and S. Johnson, and J. Kiefer referred to in the discussion of 17.1 at the end of Chapter 17, and

S. H. Brooks, "A comparison of maximum-seeking methods," *Oper. Res.*, vol. 7, 1959, pp. 430–457.

5.18 For a more detailed discussion of mathematical model-building, see

R. Bellman and P. Brock, "On the concept of a problem and problem-solving," *Amer. Math. Monthly*, vol. 67, 1960, pp. 119–134.

5.19 For the use of artificial maximization operators for computational and analytic problems, see

R. Bellman, "Functional equations in the theory of dynamic programming —V: positivity and quasilinearity," *Proc. Nat. Acad. Sci. USA,* vol. 41, 1955, pp. 743–746.

R. Kalaba, "On nonlinear differential equations, the maximum operation, and monotone convergence," *J. Math. and Mech.,* vol. 8, 1959, pp. 519–574.

R. Bellman, I. Glicksberg, and O. Gross, "Some non-classical problems in the calculus of variations," *Proc. Amer. Math. Soc.,* vol. 7, 1956, pp. 87–94.

and the discussion in Chapter 12.

5.20 For further discussion and some numerical results, see

R. Bellman, I. Cherry, and G. M. Wing, "A note on the numerical integration of a class of nonlinear hyperbolic equations," *Q. Appl. Math.,* vol. 16, 1958, pp. 181–183.

5.21 For exploitation of this quadratic behavior to derive some classical variational formulas, see

R. Bellman, "Functional equations in the theory of dynamic programming —VIII: the variation of Green's functions for the one-dimensional case," *Proc. Nat. Acad. Sci. USA,* vol. 43, 1957, pp. 839–841.

R. Bellman and H. Osborn, "Dynamic programming and the variation of Green's functions," *J. Math. and Mech.,* vol. 7, 1958, pp. 81–86.

R. Bellman and S. Lehman, "Functional equations in the theory of dynamic programming—IX: variational analysis, analytic continuation, and imbedding of operators," *Proc. Nat. Acad. Sci. USA,* vol. 44, 1958, pp. 905–907.

————, "Functional equations in the theory of dynamic programming —X: resolvents, characteristic functions and values," *Duke Math. J.,* vol. 27, 1960, pp. 55–70.

CHAPTER VI

THE LAGRANGE MULTIPLIER

> When we reflect on this struggle we may console
> ourselves with the full belief that the war of
> nature is not incessant, that no fear is felt, that
> death is generally prompt, and that the vigorous,
> the healthy, and the happy survive and multiply.
>
> CHARLES DARWIN
> *"Origin of Species,"* Chapter III.

6.1 Introduction

In the previous chapters we discussed the way in which an increase in the number of dimensions effectively interposes a barrier to the *routine* solution of variational problems by means of the techniques of dynamic programming. As we pointed out, control problems involving one state variable can be treated in a very simple fashion and require a negligible time. Questions involving two state variables are within the power of modern digital computers, but can require computing times of the order of magnitude of ten or twenty hours. Questions involving three state variables can be treated on a few machines now available, and will be amenable to a number of machines that are now in the planning or production stage, but may require even longer amounts of time.

Barring any unforeseen developments of a radical nature, we must, however, acknowledge the fact that at no time in the foreseeable future do we expect to possess machines that will handle problems involving ten or twenty state variables in any prosaic fashion.

It follows that if we wish to treat a variety of significant problems that can be formulated only in terms of functions of many state variables, we must resign ourselves to the prospect of developing a "tool chest" of special methods, each applicable to a particular range of problems. Although the desirability of uniform techniques cannot be sufficiently overemphasized, we really cannot expect to find universal solvents. In any case, the mathematician would then find life unbearably tedious.

In this chapter, and again in Chapter XVIII, we wish to present a number of techniques which appear promising and which have already been used with greater and lesser success. These combine both computational and analytic aspects, in the sense that, as always, computational advances require new analytic techniques and ingenuity.

6.2 Integral Constraints

To set the stage for the first subterfuge we shall employ, let us discuss a type of constraint we have not previously considered, an *integral constraint*.

Consider the problem of minimizing the integral

$$(6.1) \qquad J(v) = \int_0^T F(u, v)\, dt$$

subject to the differential relation

$$(6.2) \qquad \frac{du}{dt} = G(u, v), \quad u(0) = c,$$

and the integral relation

$$(6.3) \qquad \int_0^T H(u, v)\, dt = k.$$

Here u and v are to be considered to be scalar variables.

Constraints of this type enter in a very natural way if we have limited control resources. For example, we may have a fixed quantity of energy available, or, more specifically, a fixed quantity of fuel, if we are thinking in terms of trajectory problems.

To treat a problem of this type by means of dynamic programming techniques, we note that the value of the minimum now depends upon c, k, and T. Hence, we introduce the function of three variables

$$(6.4) \qquad f(c, k, T) = \underset{v}{\mathrm{Min}}\, J(v).$$

It is easy to see, using the methods of the previous chapters, that this function satisfies, to terms in $0(\Delta^2)$, the functional equation

$$(6.5)\ f(c, k, T) = \underset{v}{\mathrm{Min}}\, [F(c, v)\Delta + f[c + G(c, v)\Delta,\, k - H(c, v)\Delta,\, T - \Delta]].$$

In the limit as $\Delta \to 0$, this reduces to the quasilinear partial differential equation

$$(6.6) \qquad \frac{\partial f}{\partial T} = \underset{v}{\mathrm{Min}} \left[F(c, v) + G(c, v)\frac{\partial f}{\partial c} - H(c, v)\frac{\partial f}{\partial k} \right].$$

Whether we integrate this equation numerically, or use the discrete version of the process and the recurrence relation in (6.5), we face the prospect of dealing with sequences of functions of two variables.

As noted above, problems of this type can be handled in a routine fashion. Let us, however, see if we can devise a scheme for reducing the dimensions of the problem. What is merely a luxury in this case, where two state variables exist, will be a necessity as soon as we face more realistic control processes, involving three or more variables.

6.3 Lagrange Multiplier

Our aim is to find a way of converting a problem involving sequences of functions of two state variables into one involving sequences of functions of one variable. We shall accomplish this by invoking the classical *Lagrange multiplier*. Generally, this will enable us in a number of cases to reduce N-dimensional problems to $(N - k)$-dimensional problems, at the cost of additional computing time. We are trading *time* for *memory*.

As we shall see below, our concept of the Lagrange multiplier will be quite different from the usual one which is closely tied to the methods of calculus. This broader view enables us to handle larger classes of problems, discrete as well as continuous.

Prior to any rigorous discussion, let us proceed purely formally in a way which is familiar from the corresponding procedure that is employed in calculus problems involving functions of a finite number of variables.

In place of the original variational problem, as posed in (6.1)–(6.3), let us consider the problem of minimizing the new functional

$$(6.7) \qquad J_1(v) = \int_0^T F(u, v) \, dt + \lambda \int_0^T H(u, v) \, dt$$

subject only to the differential relation

$$(6.8) \qquad \frac{du}{dt} = G(u, v), \quad u(0) = c.$$

Here λ is an unknown parameter to be determined in a way we shall describe below.

The classical approach is to write down the Euler equation corresponding to the functional above, and in the process of solving this, to determine u and v as functions of λ. Substituting these expressions in the integral constraint of (6.3) we obtain an equation for λ, which theoretically disposes of the problem.

In actuality, of course, we face all the difficulties we have emphasized in some detail in the foregoing pages, constraints, relative minima, etc., and to be sure, even further difficulties due to the presence of the parameter λ whose value is unknown.

Consequently, let us pursue the dynamic programming approach. The important point to observe is that for any fixed value of λ, the variational problem is one that can be treated in terms of sequences of functions of *one* variable, which in this case is c. Our computational approach will then be the following. We allow λ to assume a set of values $\lambda_1, \lambda_2, \ldots, \lambda_k$. For each fixed value of λ, say λ_i, we resolve the variational problem by means of the functional equation technique of dynamic programming. Using this solution, we compute the integral

$$(6.9) \qquad \int_0^T H(u, v) \, dt = \int_0^T H(u(t, \lambda_i), v(t, \lambda_i)) \, dt = k_i.$$

In this way, as we show below, we obtain the solution of the original variational problem for a set of k-values—precisely what we wished to do in the first place, recalling the discussion in 5.18.

6.4 Discussion

Using the foregoing method, we obtain the solution directly as a function of three variables c, T, and λ, rather than a function of the three variables c, T, and k, although, as indicated, we have a means of transforming from one set of variables to another.

Fortunately, the parameter λ which appears to enter in a purely formal way, as an analytic device, has quite significant physical interpretations. In mathematical economics, it plays the role of a *"price,"* and it has a similar interpretation in many control processes. It trades *constraint value* for *return value*, the usual concept of a price.

The importance of this observation lies in the fact that in choosing values of λ to use in our computation, we can often rely upon intuition and experience to guide us initially. The solution of a particular problem involving a particular choice of k can often be simplified by using the search techniques mentioned in the previous chapter.

6.5 Several Constraints

When treating problems involving several constraints, we have our choice of several approaches. Consider the problem posed above in which we add another constraint of the form

$$(6.10) \qquad \int_0^T H_1(u, v) \, dt \le k_1.$$

The inequality formulation is more sensible when more constraints occur, since there may be no solution if equality is required.

We now have our choice of three variational problems, all susceptible to the dynamic programming treatment:

I. Minimize

$$J(v) = \int_0^T F(u, v) \, dt$$

103

subject to

(a)
$$\frac{du}{dt} = G(u, v), \quad u(0) = c,$$

(b)
$$\int_0^T H(u, v) \, dt \leq k,$$

(c)
$$\int_0^T H_1(u, v) \, dt \leq k_1.$$

II. Minimize

$$J(v) = \int_0^T F(u, v) \, dt + \lambda \int_0^T H(u, v) \, dt$$

subject to

(a)
$$\frac{du}{dt} = G(u, v), \quad u(0) = c,$$

(b)
$$\int_0^T H_1(u, v) \, dt \leq k_1.$$

III. Minimize

$$J(v) = \int_0^T F(u, v) \, dt + \lambda \int_0^T H(u, v) \, dt + \lambda_1 \int_0^T H_1(u, v) \, dt$$

subject to

(a)
$$\frac{du}{dt} = G(u, v), \quad u(0) = c.$$

6.6 Discussion

Which version of the problem we treat depends upon a number of factors, *physical*, *mathematical*, and *economic*.

In the first place, it may be true that the original control process that gave rise to the analytic problem is more meaningful in terms of one set of parameters rather than another. Furthermore, as mentioned above, we may have more of a background of experience concerning the choice of values of the parameters in one formulation.

From the mathematical point of view, it may turn out that the variational problem possesses certain desirable features in one formulation, and certain undesirable features in terms of another.

From the economic point of view, we must take account of computing time and the type of digital computer available. It may very well be that limited memory will force us to use the third formulation leading to sequences of functions of one variable rather than a routine computation of terms of sequences of functions of three variables, which the first formulation yields.

Generally speaking, the arithmetic of computers is such that it is preferable to do a hundred, or perhaps even a thousand, problems involving functions of one variable, rather than one problem involving functions of two variables.

6.7 Motivation for the Lagrange Multiplier

Before discussing the more fundamental interpretation of the Lagrange multiplier, let us outline the classical origin. Consider the finite-dimensional problem of minimizing the function of two variables $F(x_1, x_2)$ subject to the constraint that x_1 and x_2 lie on the curve $G(x_1, x_2) = 1$. Let $\bar{x}_1, \bar{x}_2$ be an extremum and set

$$(6.11) \qquad x_1 = \bar{x}_1 + y_1,$$
$$x_2 = \bar{x}_2 + y_2,$$

where y_1 and y_2 are infinitesimals. From the fact that the first variation must vanish, we obtain the equation

$$(6.12) \qquad y_1 \frac{\partial F}{\partial x_1} + y_2 \frac{\partial F}{\partial x_2} = 0,$$

and from the fact that we are constrained to the curve $G(x_1, x_2) = 1$, we derive the relation

$$(6.13) \qquad y_1 \frac{\partial G}{\partial x_1} + y_2 \frac{\partial G}{\partial x_2} = 0.$$

We have dropped the bars over the x_i to simplify the notation.

The simultaneous existence of these two relations for all y_i requires that the determinant of the y_i vanish. Thus

$$(6.14) \qquad \begin{vmatrix} \dfrac{\partial F}{\partial x_1} & \dfrac{\partial F}{\partial x_2} \\[2mm] \dfrac{\partial G}{\partial x_1} & \dfrac{\partial G}{\partial x_2} \end{vmatrix} = 0.$$

This relation, however, implies that there exists a constant with the property that the two equations

$$(6.15) \qquad \frac{\partial F}{\partial x_1} + \lambda \frac{\partial G}{\partial x_1} = 0,$$
$$\frac{\partial F}{\partial x_2} + \lambda \frac{\partial G}{\partial x_2} = 0,$$

are consistent.

Observe that these are precisely the equations that we would have obtained had we started out with the problem of determining the stationary points of the function

$$(6.16) \qquad F(x_1, x_2) + \lambda G(x_1, x_2)$$

with *no* constraints at all on the x_i.

This then is the conventional motivation of the *Lagrange multiplier* formalism. As might be expected, a rigorization of the method is not trivial, even in the finite-dimensional case, and in the calculus of variations it requires a major effort.

We shall follow a different approach in what follows, one which permits us to employ the Lagrange multiplier formalism in problems involving minimization over discrete sets, and in many cases where any derivation along the preceding lines would be most onerous.

6.8 Geometric Motivation

Let us present, in an intuitive fashion, as usual, the geometric origin of the Lagrange multiplier. The problem of maximizing a functional of the form

$$(6.17) \qquad J(u) = \int_0^T G(u, u') \, dt$$

subject to an integral constraint of the form

$$(6.18) \qquad \int_0^T H(u, u') \, dt \leq k,$$

(where, for the sake of simplicity, we have used the simplest possible differential constraint, namely $du/dt = v$), is one that can be thought of in the following terms.

Let u range over the set of all admissible functions, and compute the values of the two functionals

$$(6.19) \qquad x_1 = \int_0^T G(u, u') \, dt,$$

$$x_2 = \int_0^T H(u, u') \, dt.$$

This yields a transformation from a *function space*, the space of admissible functions u, to a subset of *two-dimensional space*, the set of points (x_1, x_2) obtained in the foregoing manner.

A set of points of the type (x_1, x_2) is often called a *"moment space."* The simplest, and most important, spaces of this type are the $(2N + 1)$-dimensional sets furnished by the Fourier coefficients of a function $u(t)$ defined over the finite interval $[0, 1]$,

$$(6.20) \qquad a_n = \int_0^1 u(t) e^{-2\pi i n t} \, dt, \quad n = 0, \pm 1, \pm 2, \ldots, \pm N,$$

where $u(t)$ is allowed to range over a function space such as the set of functions uniformly bounded over $[0, 1]$, the set of functions for which $\int_0^1 u^2(t) \, dt \leq 1$, and so on. Techniques developed by Carathéodory and others, for the special case of Fourier coefficients can be applied to far more general problems.

In order to simplify our discussion, and to understand the power of the simple geometric method we shall present, let us assume that the set of points (x_1, x_2) that we obtain in this way is a convex set in the (x_1, x_2)-plane. We may think of this set as the interior of an oval, together with its boundary.

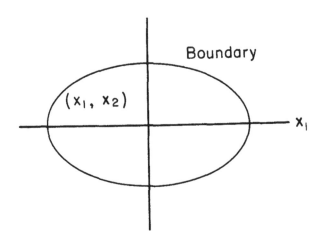

Figure 16

To solve the original maximization problem, requiring the extreme value of x_1 when x_2 has a fixed value, we must determine points on the boundary of the oval. Hence, the original maximization problem is equivalent to the problem of the determination of the boundary of our oval.

This boundary can be determined—theoretically that is—in the following fashion. We take a line, say an equation of the form

(6.21) $$ax_1 + bx_2 = k,$$

and move it parallel to itself until it is tangent to the oval. The points of tangency are precisely the boundary points, as indicated in the diagram on p. 108.

As a and b assume all values, which is to say, as we take lines in all directions, we sweep out the boundary. Observe that we are once again exploiting the *duality* of geometric figures, a locus of points is an envelope of tangents.

In order to use this information to obtain analytic results, we observe that the points of tangency are also specified by the condition that the distance of the line from the origin has an extreme value, a maximum or a minimum subject to the constraint that the line intersect the curve in at least one point. For fixed a and b, the distance from the origin is proportional

107

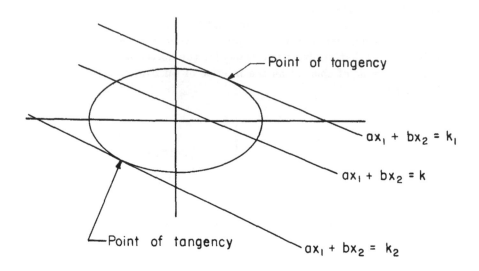

Figure 17

to k. Hence, maximization or minimization of the quantity k, which is equal to

$$(6.22) \qquad ax_1 + bx_2 = a \int_0^T G(u, u')\, dt + b \int_0^T H(u, u')\, dt,$$

yields boundary points.

If a is not equal to zero, we can divide through by a and consider the problem of determining the extreme values of

$$(6.23) \qquad \int_0^T G(u, u')\, dt + \lambda \int_0^T H(u, u')\, dt,$$

where we have set $b/a = \lambda$. We see then the "true" origin of the Lagrange multiplier.

Although the ideas are quite simple, some of the rigorous details are nontrivial. Consequently, we shall content ourselves with pointing out some results which can be obtained easily.

6.9 Equivalence of Solution

One of the ways in which we can establish the worth of this method without ensnaring ourselves in troublesome matters of rigor is to prove the following intuitive result.

"Let $u(t)$ be a function yielding the absolute minimum of the functional

$$(6.24) \qquad J(u) = \int_0^T G(u, u')\, dt + \lambda \int_0^T H(u, u')\, dt.$$

108

Let

(6.25)
$$\int_0^T H(u, u') \, dt = k,$$

for this function u(t). Then u(t) yields the absolute minimum of the functional

(6.26)
$$K(u) = \int_0^T G(u, u') \, dt$$

subject to the constraint

(6.27)
$$\int_0^T H(u, u') \, dt = k.''$$

The proof is simple and proceeds by contradiction. Let $v(t)$ be a function with the property

(6.28)
$$\int_0^T G(v, v') \, dt < \int_0^T G(u, u') \, dt,$$

while satisfying the condition

(6.29)
$$\int_0^T H(v, v') \, dt = k.$$

Then

(6.30)
$$\int_0^T G(v, v') \, dt + \lambda \int_0^T H(v, v') \, dt = \int_0^T G(v, v') \, dt + \lambda k$$
$$< \int_0^T G(u, u') \, dt + \lambda k = \int_0^T G(u, u') \, dt + \lambda \int_0^T H(u, u') \, dt,$$

a contradiction!

6.10 Discussion

The importance of the preceding result lies in the fact that it enables us to by-pass rigorous difficulties by direct experimentation in any particular problem.

Taking a particular value of λ, we are led to a particular choice of k by means of the dynamic programming solution. We know from the previous discussion that we have solved the original variational problem for this value of k. Consequently, by running through a range of λ-values, we obtain solutions for a range of k-values. If this range of values includes the values of interest to us, we have the desired solution.

If we find that no range of λ-values yields a value of k in the appropriate range, then a closer examination is required.

In many cases, the physical nature of the problem yields an intuitive monotone relation between λ and k. This is quite important, since it enables us to utilize systematic search techniques to narrow down the search for appropriate values of λ. In some cases, which we shall not discuss here, this monotonic behavior can be established rigorously.

Bibliography and Discussion

6.1 This method was first presented in

R. Bellman, "Dynamic programming and Lagrange multipliers," *Proc. Nat. Acad. Sci. USA*, vol. 42, 1956, pp. 767–769.

A variety of numerical applications will be found in

R. Bellman and S. Dreyfus, *Computational Aspects of Dynamic Programming*, Princeton University Press, Princeton, New Jersey, 1961.

6.2 For discussions of the Lagrange multiplier as a price and its role in mathematical economics, see

P. Samuelson, *Foundations of Economic Analysis*, Harvard Univ. Press, 1948.

H. Kuhn and A. W. Tucker, "Nonlinear programming," *Second Berkeley Symposium on Mathematical Statistics and Probability*, Univ. of California Press, 1951, pp. 481–492.

6.8 For further discussion of the classical background, and many references, see

E. F. Beckenbach and R. Bellman, *Inequalities*, Ergebnisse der Math., 1961.

6.9 A detailed discussion of the equivalence of different types of variational problems, with and without Lagrange multipliers, may be found in

R. Courant and D. Hilbert, *Methoden der Mathematischen Physik*, vol. 1, Interscience, 1943.

6.10 Proof of monotonicity under various hypotheses may be found in

R. Bellman, I. Glicksberg, and O. Gross, *Some Aspects of the Mathematical Theory of Control Processes*, The RAND Corporation, Report R-313, 1958,

and in the book cited above by Bellman and Dreyfus.

The equation in (6.6) can be used to show the connection between the Lagrange multiplier and the "price," namely $\partial f/\partial k$. Compare this equation with the functional equation obtained from the problem of minimizing

$$J(v) = \int_0^T [F(u, v) - \lambda H(u, v)] \, dt,$$

subject to the differential equation (6.2). Then if we set $f(c, T) = \text{Min } J(v)$, we see that

$$\frac{\partial f}{\partial T} = \underset{v}{\text{Min}} \left[F(c, v) - \lambda H(c, v) + G(c, v) \frac{\partial f}{\partial c} \right].$$

Comparing this equation with (6.6), we see that a relation exists between λ and $\partial f/\partial k$.

For a closely related discussion, see

S. Dreyfus, *Dynamic Programming and the Calculus of Variations*, Jour. Mathematical Analysis and Applications, 1960.

See also the discussion on page 52 of

R. Bellman, I. Glicksberg, and O. Gross, *Some Aspects of the Mathematical Theory of Control Processes*, The RAND Corporation, Report R-313, January 16, 1958.

CHAPTER VII

TWO-POINT BOUNDARY
VALUE PROBLEMS

There was a most ingenious architect who had
contrived a new method for building houses, by
beginning at the roof, and working downward
to the foundation. . . .

JONATHAN SWIFT
"Gulliver's Travels," Voyage to Laputa.

7.1 Introduction

In previous chapters we have discussed the problem of solving variational
questions of the following form:
"Minimize

$$(7.1) \qquad J(y) = \int_0^T g(x, y) \, dt$$

subject to the constraints

$$(7.2) \qquad \frac{dx}{dt} = h(x, y), \quad x(0) = c."$$

As we know, classical variational techniques yield a two-point boundary
value problem. The Euler equation is tied down by the prescribed condition at $t = 0$, and by a condition at $t = T$ which is obtained from the
variational process itself; cf. the results of 1.12.

From the standpoint of the theory of control processes, a problem of the
foregoing type corresponds to a situation in which a system has a prescribed
initial state, but is not subject to any terminal conditions. In many
important instances, such as optimal trajectory and space travel, however,
there are prescribed terminal conditions as well.

Here, we wish to consider some of the details of the computational
solution when we have terminal conditions of the form

$$(7.3) \quad (a) \qquad\qquad x(T) = c_1, \quad \text{or}$$

$$(b) \qquad\qquad x'(T) = c_2,$$

111

or, more generally, a relation of the form

(7.4)
$$k[x(T),\, x'(T)] = 0.$$

We shall also discuss the case where there exist constraints of the form $|x'(t)| \leq k$. Similar methods can be applied to treat the case where there are relations fixing the values of x at internal points of the interval.

Finally, we shall show how functional equation techniques can be applied to one of the most important types of two-point boundary value problems, that of determining the principal characteristic value of a Sturm-Liouville equation of the form

(7.5)
$$u'' + \lambda\phi(t)u = 0,$$
$$u(0) = u(1) = 0.$$

As in all of our previous chapters, we are interested only in the formalism at the moment, and we will avoid all rigorous details of any complexity. With some effort, these methods can be demonstrated to be rigorously valid.

7.2 Two-point Boundary Value Problems

Let us briefly review some comments we have made above. Given a second-order differential equation of the form

(7.6)
$$u'' + h(u, u') = 0,$$

where the function $h(u, u')$ appearing on the right-hand side satisfies mild conditions (but more than just continuity), we know there is a unique solution $u(t)$ in some interval $[0, t_0]$ determined by initial conditions of the form

(7.7)
$$u(0) = a_1, \quad u'(0) = a_2.$$

By use of a computing machine of one type or another, an analogue computer in special cases, and a digital computer in general, we can obtain the numerical solution of this equation to practically any desired order of accuracy.

If, however, the conditions on the solution are of the form

(7.8)
$$u(0) = c_1, \quad k[u(T), u'(T)] = 0,$$

a two-point condition, then we have to face some problems that were present in the one-point case, but rather subdued.

In the first place, uniqueness is not just mathematical frippery in this case, but a very real problem. We have previously shown (in 1.16) how quite simple and sensible problems in the calculus of variations can give rise to equations with a denumerable set of solutions satisfying the same end-conditions. It is thus completely reasonable to expect that boundary conditions of the type appearing in (7.8) define a set of solutions, rather than a particular solution.

Secondly, even after we have settled upon the precise member of the class of solutions that we wish to compute, we still face a grave problem of accuracy. The usual technique used to solve a problem in which the conditions are as given in (7.8) is to guess an initial slope, $u'(0)$, and then integrate the equation over the interval $[0, T]$ in any of a number of conventional manners. Once this has been done, the value of $u(T)$ obtained in this way is compared with the required value, and, generally, we compute the required function of $u(T)$ and $u'(T)$. If the agreement is sufficiently close, we accept this approximation to the solution; if not, we choose a new value of $u'(0)$. This is not a particularly desirable procedure, and, in practice a number of disagreeable things can happen.

If the Euler equation is linear, life is very much simpler. However, even in the case of linearity, if the dimension of the system is large, we face problems of some complexity.

7.3 Applications of Dynamic Programming Techniques

If (7.6) is the Euler equation associated with a variational problem of the type:
"Minimize

$$(7.9) \qquad \int_0^T g(u, u') \, dt$$

subject to the conditions

$$(7.10) \qquad u(0) = c_1, \quad k[u(T), u'(T)] = 0";$$

then we can apply the functional equation technique of dynamic programming to treat the problem as an initial value problem. Furthermore, we now have a means of determining bounds on the accuracy of the solution.

Due to the fact that so much of mathematical physics is dominated by teleological principles, there is a high probability that a differential equation that arises in the course of an investigation can be derived from a variational principle.

7.4 Fixed Terminal State

Since it appears rather magical at first sight that the functional equation technique does allow us to replace two-point problems by initial problems, let us discuss in some detail what is involved in the computational solution of problems of this type.

Let us begin with the simplest case of a problem requiring that the system start in a fixed initial state and end up in a fixed terminal state. Consider the problem of minimizing the integral

$$(7.11) \qquad J(u) = \int_0^T g(u, u') \, dt$$

113

subject to the two conditions

(7.12) $$u(0) = c_1, \quad u(T) = c_2.$$

As before, we replace the continuous variation by variation over a discrete set. Proceeding as in 4.11 we obtain a recurrence relation of the form

(7.13) $$f_{N+1}(c) = \underset{v}{\text{Min}} \, [g(c, v) + f_N(c + v\Delta)],$$

with the proviso that the final state of the system must be the assigned value c_2.

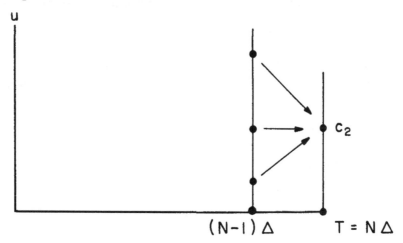

Figure 18

What this means in effect is that at the last stage of the process (when there is one decision remaining), regardless of the value of u, which is to say, regardless of the state of the system, the choice of the slope function must be such as to force the solution to c_2.

Consequently, the terminal constraint fixes the function $f_1(c_1)$ by the relation

(7.14) $$f_1(c) = g(c, v),$$

where

(7.15) $$v = \frac{c_2 - c}{\Delta}.$$

Hence,

(7.16) $$f_1(c) = g\left(c, \frac{c_2 - c}{\Delta}\right).$$

114

Here, c_2 is taken to be a fixed quantity and c is considered to be the variable quantity. As long as we have a free choice of v, we can accomplish this task.

We observe once again the fact that constraints simplify the solution, in the sense that $f_1(c)$ is automatically determined without a minimization. Having determined $f_1(c)$, the remaining elements of the sequence $\{f_k(c)\}$ are determined by means of (7.13) with no further reference to the terminal condition.

7.5 Fixed Terminal State and Constraint

Suppose that we now add the condition that $u'(t)$ be bounded in absolute value, i.e.

$$(7.17) \qquad |u'(t)| \leq k.$$

Then it need no longer be possible for c_2 to be reached from *all* initial states, In particular, we see that with one stage remaining, we must have

$$(7.18) \qquad \left| \frac{c_2 - c}{\Delta} \right| \leq k.$$

Hence,

$$(7.19) \qquad c_2 - k\Delta \leq c \leq c_2 + k\Delta.$$

Thus, $f_1(c)$ is defined only for this set of values of c, $f_2(c)$ is defined for a slightly larger set, and so on.

Once again, the net result of the constraint is beneficial in two ways. As noted previously, it greatly simplifies the search process for a solution, and here, furthermore, it simultaneously reduces the number of values which must be tabulated at each stage. Hence, the presence of the constraint greatly simplifies the process of obtaining the solution—provided that we employ dynamic programming techniques.

7.6 Fixed Terminal Set

Sometimes, in place of requiring that the system be in a fixed state at the end of the process, we require merely that it be in one of a given set of states, $c_{21}, c_{22}, \ldots, c_{2M}$. Then in place of (7.16) we would have the relation

$$(7.20) \qquad f_1(c_1) = \operatorname*{Min}_{c_{2i}} g\left(c_1, \frac{c_{2i} - c_1}{\Delta}\right).$$

From this point on, the equation of (7.13) takes over.

Observe the interesting fact that the functional equation governing the process is the same, regardless of the terminal conditions. This is not really surprising, since precisely the same situation exists in the classical theory, where the analytic form of the Euler equation is independent of the end conditions.

7.7 Internal Conditions

Let us briefly discuss the application of the same techniques to the problem of minimizing the integral

(7.21)
$$J(u) = \int_0^T g(u, u')\, dt,$$

subject to the constraints

(7.22)
$$u(0) = c_1,$$
$$u(T) = c_2,$$
$$h[u(s), u'(s)] = 0,$$

where s is some fixed value in $[0, T]$. As above, we introduce the function

(7.23)
$$f(c_1, T) = \operatorname*{Min}_u J(u).$$

For $T > s$, the equations are as before, except that at $T = s + \Delta$, the choice of u' is fixed by the condition $h[u(s), u'(s)] = 0$. For $T \le s$, the equations are also as before.

7.8 Characteristic Value Problems

Let us now indicate how functional equation techniques may be applied to the problem of finding the smallest characteristic value of the Sturm-Liouville equation

(7.24)
$$u'' + \lambda\phi(t)u = 0,$$
$$u(0) = u(1) = 0.$$

Our assumption will be that $\phi(t)$ is a continuous, uniformly positive function over $[0, 1]$. It is easy to see that this equation is obtained from the problem of determining the minimum of the ratio

(7.25)
$$J(u) = \frac{\displaystyle\int_0^1 u'^2\, dt}{\displaystyle\int_0^1 \phi(t)u^2\, dt},$$

subject to the constraints $u(0) = u(1) = 0$, or, equivalently, from the problem of determining the minimum of

(7.26)
$$\int_0^1 u'^2\, dt,$$

subject to the constraints

(7.27)
$$\int_0^1 \phi(t)u^2\, dt = 1,$$
$$u(0) = u(1) = 0.$$

This is a variational problem of the kind that we have already met, involving an integral constraint.

Thus, in place of the original problem, we consider the more general problem of minimizing the integral

$$(7.28) \qquad J(u) = \int_a^1 u'^2 \, dt,$$

subject to the constraints

$$(7.29) \quad (a) \qquad \int_a^1 \phi(t)u^2 \, dt = 1,$$

$$(b) \qquad u(a) = c,$$

$$(c) \qquad u(1) = 0.$$

Introducing the function

$$(7.30) \qquad f(a, c) = \underset{u}{\text{Min}} \int_a^1 u'^2 \, dt,$$

there is now no difficulty in deriving a functional equation for $f(a, c)$ which can be used to yield a partial differential equation or a discrete recurrence relation for computational purposes.

Observe, however, that the initial condition $f(1, c) = 0$ is no longer valid. The constraint of (7.29a) forces $f(a, c)$ to have a singular behavior as $a \to 1$. This asymptotic behavior can readily be determined by solving the problem of minimizing

$$(7.31) \qquad J(u) = \int_a^1 u'^2 \, dt$$

subject to the constraints of (7.29b) and (7.29c), and

$$(7.32) \qquad \phi(1) \int_a^1 u^2 \, dt = 1.$$

If $\phi(1) = 0$, then we replace (7.32) by

$$(7.33) \qquad \phi(a) \int_a^1 u^2 \, dt = 1.$$

Bibliography and Discussion

7.1 For a discussion of numerical treatment of Sturm-Liouville problems by other means, see,

L. Collatz, *Numerische Behandelung von Differentialgleichungen*, Z. Aufl., Berlin, 1955.

and

R. Bellman, "On the determination of characteristic values for a class of Sturm-Liouville problems," *Illinois J. Math.*, vol. 2, 1958, pp. 577–585.

7.8 Further details may be found in Chapter 8 of

R. Bellman, *Dynamic Programming*, Princeton Univ. Press, Princeton, N.J. 1957.

As mentioned above, an entirely different approach to two-point boundary value problems may be found in

R. Kalaba, "On nonlinear differential equations, the maximum operation and monotone convergence," *J. Math. and Mech.*, vol. 8, 1959, pp. 519–574.

Some of the essential ideas of this paper are given in Chapter 12.

CHAPTER VIII

SEQUENTIAL MACHINES AND THE SYNTHESIS OF LOGICAL SYSTEMS

> Just as virtue is not more virtuous for having
> flourished longer, so I hold a truth is no wiser for
> being older.
> MICHEL DE MONTAIGNE
>
> *"The Autobiography of Michel de Montaigne,"*
> Chapter XXIII

8.1 Introduction

In the preceding chapters, we have used mathematical models of physical control systems to motivate our discussion. In this chapter—the last one to be devoted to deterministic control processes—let us turn to a more abstract setting and spend some time on general logical systems.

We shall first treat some "detection" or "diagnosis" problems associated with the concept of a sequential machine, and then turn to the study of the optimal synthesis of logical systems designed for certain tasks. At the end of the chapter there will be found some references indicating the connections between these problems and some aspects of medical diagnosis.

8.2 Sequential Machines

We shall use the term *sequential machine* to signify a mathematical system consisting of a finite set of quantities, $I_1, I_2, \ldots, I_M$, which we call *inputs*, a finite set of quantities, $O_1, O_2, \ldots, O_N$, which we call *outputs*, and a finite set of quantities $x_1, x_2, \ldots, x_K$, which we call the *states* of the system.

These quantities are interrelated by means of the following Markovian properties:

(8.1) (a) The present state of the system depends only upon the past state and the past input.

    (b) The present output of the system depends only upon the present state and the present input.

We then have two "multiplication tables" describing the future states and outputs, given the past states and inputs:

Present Input \	Present State			
	x_1	x_2	$\cdots$	x_N
I_1	x_{11}	x_{12}		x_{1N}
I_2	$\cdot$	$\cdot$		$\cdot$
	$\cdot$	$\cdot$		$\cdot$
$\cdot$				
$\cdot$				Next State
I_M	x_{M1}	x_{M2}	$\cdots$	x_{MN}

with a similar table for the outputs, which we shall call O_{ij}.

Present Input \	Present State			
	x_1	x_2	$\cdots$	x_N
I_1	O_{11}	O_{12}		O_{1N}
I_2	$\cdot$	$\cdot$		$\cdot$
$\cdot$	$\cdot$	$\cdot$		$\cdot$
$\cdot$				Output
I_M	O_{M1}	O_{M2}	$\cdots$	O_{MN}

A fundamental problem in this field is that of using this information to determine the current state of a sequential machine, assumed constant until an input is applied. One part of this problem is that of determining when this is possible, and another part is that of determining the state in some optimal fashion, i.e. as quickly as possible, as cheaply as possible, and so on, when this can be done.

We wish to show that the theory of dynamic programming can be utilized to provide a conceptual and analytic basis for problems of this nature, and, in addition, to furnish computational algorithms. As a simple example of these techniques, we shall consider the classical coin-weighing problem, which has by now been solved in a large number of different ways.

8.3 Information Pattern

Let us now consider the formulation of the problem of determining optimal testing methods within the framework of the theory of dynamic programming. To begin with, we must introduce state variables which describe the state of our knowledge at any stage of the process. This is usually one of the most difficult aspects of these problems.

Suppose that initially it is known that the system is in one of a number of possible states, $x_{a1}, x_{a2}, \ldots, x_{ak}$. The set

$$(8.2) \qquad S = [x_{a1}, x_{a2}, \ldots, x_{ak}]$$

is called the *information pattern*, a concept we shall meet again in later chapters (e.g. 16.2). Conceivably, S may be the set of all possible x_i, which we call Σ.

As we insert inputs and observe outputs, the information pattern, which is to say, the set of possible current states of the machine, will change.

8.4 Ambiguity

There are now two possibilities. Either the structure of the multiplication tables given above is such that by means of some appropriate sequence of inputs we can finally reach a point where we definitely know the state of the machine, or we cannot. Note that there are several different problems that can be treated at this point. One is that of performing a sequence of tests to determine the *initial* state of the machine, another is that which we have already proposed. We shall consider only the former problem.

If we cannot ever come to a definite conclusion, it is of interest to determine the minimum uncertainty attainable, measured perhaps in terms of the minimum number of possible states of the machine. If we can reach a definite conclusion, it is of interest to determine the least number of inputs required to attain certainty.

In what follows, we wish to introduce the concept of *ambiguity*, and show how to calculate it by means of functional equations.

8.5 Functional Equations

Given a set of elements S, we introduce the scalar function

$$(8.3) \qquad n(S) = \text{the number of elements in } S.$$

Let us call $n(S)$ the *norm* of S.

Let us now pose the following problem: "Given the information that the current state of the sequential machine is a member of a set S, we wish to insert a sequence of inputs which will minimize the norm of the set of possible states resulting after k stages."

This is to be a sequential, or feedback, process, since we allow, and as a matter of fact encourage, observation of the outputs as the testing proceeds.

Let us define

$$(8.4) \quad \alpha_0(S) = n(S),$$

and

$$(8.5) \quad \alpha_k(S) = \text{the norm of the set describing the information pattern}$$
$$\text{after } k \text{ trials, using an optimal testing policy.}$$

In order to derive a recurrence relation connecting $\alpha_k(S)$ with the function $\alpha_{k-1}(S)$, let us describe what happens when we use a particular input I. As a result of observing the output O, and of using the knowledge of the previous information pattern S, we deduce from the multiplication tables that the current state of the system must belong to a new set which we shall call S_{IO}.

Furthermore, from the multiplication table for outputs, we know that O itself must be a member of a set which is determined by I and the set S. Call this set of possible outputs $T(I, S)$. We do not, of course, know the precise output beforehand, since we do not know the precise state of the machine.

Since we do not know the precise output, we must acknowledge the possibility that it will be the worst possible from our point of view. With this in mind, the input I must be chosen so as to minimize.

Putting these remarks together, we see that the functional equation satisfied by the elements of the sequence $\{\alpha_k(S)\}$ is

$$(8.6) \qquad \alpha_k(S) = \operatorname*{Min}_I \left[\operatorname*{Max}_{O \in T(I,S)} \alpha_{k-1}(S_{IO}) \right], \quad k \geq 1.$$

This is an application of the principle of optimality to which we have repeatedly appealed in the foregoing pages.

8.6 Limiting Case

Let us now consider the limiting case obtained as we let the number of allowable tests become infinite. Since the function $\alpha_k(S)$ is clearly monotone decreasing as a function of k as k increases, we can define for each set S the function

$$(8.7) \qquad \alpha(S) = \lim_{k \to \infty} \alpha_k(S).$$

We call this function $\alpha(S)$ the *ambiguity* of S. It satisfies the functional equation

$$(8.8) \qquad \alpha(S) = \operatorname*{Min}_I \left[\operatorname*{Max}_{O \in T(I,S)} \alpha(S_{IO}) \right].$$

If $\alpha(S) = 1$ for every $S \subset \Sigma$, we say that the sequential machine is *unambiguous*.

8.7 Discussion

In passing, let us note that we have used a rather useful analytic device in discussing the problem of determining the state of the machine. Not only have we posed a *feasibility* problem, but we have associated it with a *variational* problem. Frequently, the two in combination are easier to attack than the feasibility problem alone.

8.8 Minimum Time

Along these very same lines, let us consider the problem of determining a testing problem that will reduce S to a single element as quickly as possible. This is a generalized "trajectory" problem, an abstract version of the "bang-bang" control process discussed in Chapter IV.

Define the new function as

(8.9) $f(S) = 0$, if $n(S) = 1$,

= the minimum number of trials required to reduce S to a single element, provided that $n(S) \neq 1$ initially.

If it is not possible to reduce S to a single element, then $f(S)$ will have an infinite value. The function clearly satisfies the functional equation

(8.10) $$f(S) = 1 + \underset{I}{\text{Min}} \left[\underset{O \in T(I,S)}{\text{Max}} f(S_{IO}) \right].$$

The problem of determining whether or not a given sequential machine is ambiguous or not has thus been made equivalent to solving (8.10). Although this may or may not constitute any advance, it does permit various computational approaches to be used to treat the problem.

8.9 The Coin-weighing Problem

As a simple illustration of these techniques, consider the classic puzzle of using an equal-arm balance to detect one heavy coin in a lot of N coins of similar appearance. Let

(8.11) $f_N =$ the maximum number of weighings required using an optimal policy.

A word of explanation is required as to what we mean by an optimal policy. It is always possible by means of a fortuitous choice to determine the heavy coin in one weighing. We wish, on the contrary, to prescribe a testing procedure which *guarantees* finding the heavy coin in, at most, a fixed number of weighings. It is this fixed number which we call f_N.

At each stage, we are allowed to weigh one batch of k coins against another, and observe the outcome. Two situations can arise. Either the two sets of coins will balance, or they will not. If the two sets balance, the heavy coin must be one of the remaining $N - 2k$ coins; if they do not balance, we know that the heavy coin is in one of the sets of k coins. Since we must take account of both contingencies, we obtain the relation

(8.12) $$f_N = 1 + \underset{\leq k \leq N/2}{\text{Min}} \ \text{Max}[f_k, f_{N-2k}].$$

To minimize, we clearly want to make k and $N - 2k$ as equal as possible.

Consequently, we take $k = [N/3]$ or $[N/3] + 1$, depending upon whether N has the form $3m + 1$ or $3m + 2$. This yields one of the well known solutions.

8.10 Synthesis of Logical Systems

Another rather interesting class of generalized trajectory problems of the type described above occurs in connection with the design of logical systems, which in turn is intimately connected with the design of computers.

Consider an input signal of the form

$$(8.13) \qquad x = \begin{bmatrix} x_1 \\ x_2 \\ \cdot \\ \cdot \\ \cdot \\ x_M \end{bmatrix},$$

an M-dimensional vector, where the components x_i assume only the values 0 or 1. This signal is to be fed into a series of components (black boxes), which transform a vector of this type into a vector of similar type.

As usual in the study of physical systems, there are two phases of study. We may wish to describe the operation of a particular system, or class of systems, or we may wish to construct systems with certain desirable qualities. It is the second question we propose to discuss here.

8.11 Description of Problem

The choice of a component is clearly equivalent to the choice of a transformation, a transformation which converts a particular input into a specified output. Schematically,

$$x \to \boxed{T(q_1)} \to \boxed{T(q_2)} \to \ldots \boxed{T(q_N)} \to x_N = y,$$

this being equivalent to the relations $x_1 = T(q_1, x)$, $x_2 = T(q_2, x)$, ..., $x_N = T(q_{N-1}, x_{N-1})$.

The variables q_i, as usual, denote decision variables. The choice of a q_i determines the type of component to be used at the i-th stage, and hence the transformation that is effected.

There are now several questions that can be asked:

A.　(i) With a preassigned set of components available at each stage, is it possible to construct a network which will convert a given input x into a given output y?

(ii) Assuming that networks of the desired type exist, how do we construct most efficient networks?

(iii) Assuming that at most N stages are allowable, how close can x_N come to y?

(iv) Given a set of inputs, $[x^{(1)}, x^{(2)}, \ldots, x^{(R)}]$, and a maximum of N stages, what circuit minimizes the maximum deviation from assigned outputs $[y^{(1)}, y^{(2)}, \ldots, y^{(R)}]$?

8.12 Discussion

The first problem posed in A.(i) of the preceding section is commonly called a *feasibility problem*. Quite often, problems of this type, "existence theorems," are exceedingly difficult. One way to tackle this problem is by way of the variational problems posed in (ii) and (iii), a point already made in 8.7. As we shall see, the problem in (iii) is easier to study than that in (ii), and, actually, is one way of approaching the problem posed in (ii).

8.13 Introduction of a Norm

In order to measure the deviation of x_N from y, we must introduce a metric. Let us use the norm

$$(8.14) \qquad \|y - z\| = \sum_{i=1}^{N} |y_i - z|_i.$$

Observe that for vectors of the type we are considering, this counts the number of components in y and z which are different.

8.14 Dynamic Programming Approach

Let us introduce the function

$(8.15) \quad f_N(x) =$ the minimum deviation of the output x_N from y
$$= \operatorname*{Min}_{\{q_i\}} \|x_N - y\|.$$

For $N = 1$, we have the relation

$$(8.16) \qquad f_1(x) = \operatorname*{Min}_{q_1} \|T_1(q_1, x) - y\|,$$

while for $N > 1$ we have the recurrence relation

$$(8.17) \qquad f_N(x) = \operatorname*{Min}_{q_1} f_{N-1}[T_1(q_1, x)].$$

Provided that the dimension M is not too large, we have a routine computational solution to the problem A.(iii) posed in 8.11.

8.15 Minimum Number of Stages

One way of approaching the problem of determining the minimum number of stages required to convert x into y is to use the preceding problem to ascertain the first value of N for which $f_N(x)$ is equal to zero. This is a technique we discussed above in connection with the "bang-bang" control problem.

125

Alternatively, we may introduce the function

(8.18) $g(x) =$ the minimum number of stages required to convert x into y.

Then, we have the functional equation

(8.19)
$$g(x) = \operatorname*{Min}_{q_1} g[T_1(q_1, x)].$$

This equation suffers from the defect that the unknown function appears on both sides of the equation, which means that we do not possess a routine method for calculating it. On the other hand, we can now employ various techniques of approximation, in particular "approximation in policy space," a technique mentioned above, and in more detail in Chapter XV.

8.16 Medical Diagnosis

The study of sequential machines has as an immediate application the use of digital computers for diagnostic purposes. In Chapter XI we introduce the concept of a Markovian decision process, which may yield a more realistic model.

References to previous work in this field will be found below.

Bibliography and Discussion

8.1–8.9 The discussion here follows

R. Bellman, "Sequential machines, ambiguity and dynamic programming," *J. Assoc. Comp. Mach.*, vol. 7, 1960, pp. 24–28.

The concept of a "sequential machine" was introduced by

E. F. Moore, "Gedanken-experiments on sequential machines," *Automata Studies, Annals of Math.* Series No. 34, 1956, pp. 129–153.

See also

S. Ginsburg, *On the Length of the Smallest Uniform Experiment which Distinguishes the Terminal States of a Machine*, Research Report, The National Cash Register Co., Electronics Div., Hawthorne, California, 1957.

S. Huzinu, "On some sequential machines and experiments," *Memoirs of Fac. Sci.*, Kyushu Univ., Ser. A., vol. XII, No. 2, 1958.

———, On some sequential equations, *Memoirs of Fac. Sci.*, Kyushu Univ., Ser. A, vol. XIV, No. 1, 1960, pp. 50–62.

G. H. Mealy, "A method for synthesizing sequential circuits," *Bell System Tech. J.*, vol. 34, 1955, pp. 1045–1079.

8.9 See, for example,

M. J. E. Golay, "Notes on the penny-weighing problem, lossless symbol coding with non-primes, etc.," *IRE Trans. Prof. Group on Information Theory*, vol. IT-4, 1958, pp. 103–109.

The similar problem for the case of two or more heavy coins is exceedingly difficult. For some preliminary results, see

S. S. Cairns, *Balance Scale Sorting*, the RAND Corporation, Paper P-736, September 7, 1955.

The information pattern associated with the two-coin case becomes extraordinarily complex as the process continues. A general approach to these sequential testing problems does not yet exist.

For an entirely different approach to weighing problems, predicated upon the assumption that errors will occur, see the following papers, where further references may be found.

H. Hotelling, "Some improvement in weighing and other experimental techniques," *Ann. Math. Stat.*, vol. 15, 1944, pp. 297–306.

A. M. Mood, "On Hotelling's weighing problem," *Ann. Math. Stat.*, vol. 17, 1946, pp. 432–446.

K. S. Banerjee, "Weighing designs," *Calcutta Stat. Assn. Bull.*, vol. 3, 1950–1951, pp. 64–76.

D. Raghavero, "Some optimum weighing problems," *Ann. Math. Stat.*, vol. 30, 1959, pp. 295–303.

For an expository account of the coin-weighing problem discussed in the text, see

C. A. B. Smith, *Math. Gazette*, vol. XXXI, 1947, pp. 31–39.

See also, for a discussion of a closely related problem,

T. J. Fletcher, "The *n* prisoners," *Math. Gazette*, vol. 40, 1956, pp. 98–102.

For an application of dynamic programming to the closely related problem of determining the location of a defective component in a complex system, see

B. Gluss, "An optimum policy for detecting a fault in a complex system," *Oper. Res.*, vol. 7, 1959, pp. 468–477.

See also

R. Bellman and B. Gluss, *A Policy for Isolating Defective Coins*, to appear.

M. Sobel and P. A. Groll, "Group testing to eliminate efficiently all defectives in a binomial sample," *Bell Tech. J.*, Sept. 1959.

R. Dorfman, "The detection of defective members of a large population," *Ann. Math. Stat.*, vol. 14, 1943, pp. 436–440.

W. Feller, *Probability Theory and Its Applications*, J. Wiley, New York, 1950, Ex. 18, p. 189.

P. Ungar, "The cutoff point for group testing," *Comm. Pure Appl. Math.*, vol. XIII, 1960, pp. 49–54.

S. Firstman and B. Gluss, *Search rules for automatic fault location*, RM-2514, The RAND Corporation, Jan. 1960.

For a discussion of the problem of synthesizing a reliable system by means of duplication of function of component parts, see

R. Bellman and S. Dreyfus, "Dynamic programming and the reliability of multicomponent devices," *Oper. Res.*, vol. 6, 1958, pp. 200–206.

Finally, for a discussion from a different type of view, see

N. T. Gridgeman, "Sensory item sorting," *Biometrics*, vol. 15, 1959, pp. 298–306.

where many further references may be found.

8.10–8.15 Here we follow

R. Bellman, J. Holland, and R. Kalaba, "On an application of dynamic programming to the synthesis of logical systems," *J. Assoc. Comp. Mach.*, vol. 6, 1959, pp. 486–493.

The problem of transforming a system from one state to another at minimal cost is one that arises in many parts of applied mathematics. In previous chapters, we met it under the guise of the trajectory problem and the "bang-bang" control problem. See also

R. Bellman, "On a routing problem," *Q. Appl. Math.*, vol. 16, 1958, pp. 87–90.

R. Kalaba, "On some communication network problems," *Proc. Symposium on Appl. Math.*, April, 1958.

R. Bellman and R. Kalaba, "On k-th best policies," *J. Soc. Ind. Appl. Math.*, to appear.

8.16 See

R. S. Ledley and L. B. Lusted, "Reasoning foundations of medical diagnosis," *Science*, vol. 130, 1959, pp. 9–21,

where a large number of additional references may be found.

For a bibliography of related matters, see

D. Netherwood, "A selected bibliography of logical machine design," *IRE Trans.*, PGEC, June, 1958.
(I am indebted to P. Metzelaar for this reference.)

For applications of concepts of the foregoing ideas to cancer chemotherapy and many further references, see

N. Mantel, "Principles of Chemotherapeutic Screening," *Proc. of 4th Berkeley Symposium*, Univ. of Calif., 1961.

CHAPTER IX

UNCERTAINTY AND
RANDOM PROCESSES

Persecution is used in theology, not in arithmetic,
because in arithmetic there is knowledge, but in
theology there is only opinion. So whenever you
find yourself getting angry about a difference of
opinion, be on your guard; you will probably
find, on examination, that your belief is getting
beyond what the evidence warrants.

BERTRAND RUSSELL
"Unpopular Essays"

9.1 Introduction

Although the processes we have so far considered have occasionally
forced us to employ a certain amount of mathematical sophistication and
to invade areas of some analytic complexity, conceptually they have been
on a fairly simple level. This has been a consequence of our assumption of
determinism.

By this we mean the following. Not only have we assumed that the state
of the system was fully known at each stage, but we have also supposed that
the choice of decision vector, q, led to a definite transformation of the system
from the state p to the state $T(p, q)$, where $T(p, q)$ is a predetermined
function of p and q. Furthermore, we have supposed the existence of a
definite criterion for the evaluation of the process, known in advance.

We shall begin our discussion of more complex mathematical theories
by explaining in some detail why it is necessary in many processes to
renounce the simple, elegant and comfortable analytic structure that we
have erected to house control processes. After so doing, we will turn to the
study of more elaborate frameworks. As we shall see, these will be of even
greater elegance, certainly of greater complexity, and thus, of course, all
the more fascinating as regions for fruitful research.

It must, in all justice, be admitted, however, that never again will
scientific life be as satisfying and serene as in the days when determinism
reigned supreme. In partial recompense for the tears we must shed and the

toil we must endure is the satisfaction of knowing that we are treating significant problems in a more realistic and productive fashion.

9.2 Uncertainty

There are many faces to the mask of uncertainty. We shall dwell upon one for a few chapters and then turn to even more complex aspects of the perplexing panorama of possibilities.

We begin by boldly upsetting the law of cause and effect. Let us assume, as above, that we know the state of the system both before and after the occurrence of the transformation resulting from a decision, and that we are aware of our ultimate aims at all times. However, differing from above, let us now suppose that a decision, a choice of q, does *not* imply a unique transformation, $T(p, q)$, as it did in the deterministic case. Rather, a choice of q now yields a set of possible outcomes. Having made the decision q in the state p, sometimes we will observe the resultant state p_{11}, sometimes the resultant state p_{12}, and so on. Apparently, the rigid rod of cause and effect has weakened.

In the face of a disturbing train of events of this type, we have an obvious recourse. It may well be that the explanation of this violation of a basic premise of the physical world lies not in the nature of the phenomenon itself, but rather in our incomplete description of it. Examining the state vector p, we may find that there are not enough components to yield a *complete* description of any current state of the system under observation. In a situation of this kind, it is reasonable to expect inaccurate prediction of the future behavior. If we do wish to predict the precise future behavior of the system, it is necessary to include additional information in the form of additional components of p.

To illustrate these points, consider the problem of predicting the outcome of the tossing of a coin. Since this is taken as the standard example of a situation involving uncertainty, it is particularly interesting to analyze it in some detail.

Presumably, given the initial position of the coin in phase space, the initial impulse imparted to our thumb, the velocity of the wind, the shape and elastic properties of the floor, and other pertinent information, we should be able to utilize classical mathematical and physical techniques to predict the occurrence of heads or tails, or, if the coin is sufficiently thick, the possibility of its standing on end.*

It is perhaps not superfluous to add that any research project of this type will require an enormous effort, combining experimental, theoretical, and computational programs of some magnitude, merely to collect some of the basic parameters. Having gone to all of this trouble, all will still not be

---

* It will be a challenging exercise for the reader to compute this possibility, given a non-zero thickness.

smooth sailing. The first critical observer we meet, either in the form of our own mathematical conscience or in some other guise, will point out that we are in a situation where small causes can produce great effects. In other words, a quite small change in the initial conditions of the toss can very easily change the final position from that of heads to tails. In mathematical and physical parlance, we are examining a highly *unstable* process.

It follows that we cannot proceed in any casual manner, cutting calculating corners here and there. In order to make an exact prediction, we must use precise tools.

This means that Einsteinian mechanics must be employed rather than Newtonian mechanics—a bit of embarrassment since the newer theory has not yet been successfully applied to rigid body problems. However, other even more unpleasant prospects loom ahead. Granted an at present non-existent ability to overcome this obstacle, we face the conceptual block of quantum mechanics, a theory which stubbornly refuses to admit even the possibility of ultimate precision.

The more we examine the goal of an exact solution, the more we are forced to acknowledge its unattainability due either to philosophic or practical premises. On pragmatic grounds, we are forced then to adopt an approximate description of the system, and to accept an approximate description of its future behavior.

9.3 Sour Grapes or Truth?

At this stage, rational behavior yields to a certain amount of emotional bias and wishful thinking. There are quite eminent scientists who earnestly believe that what we have invoked merely on the ground of pragmatism must actually be upheld on the grounds of inexorable physical logic. To the members of this group, determinism is merely a mathematical artifice which can be used very successfully to simplify the study of many complex processes where average effects alone are required.

Others, equally eminent, resolutely maintain that the concept of a probabilistic process (of the type we shall discuss in some detail in what follows) is itself a mathematical fiction, introduced merely to compensate for the shortcomings of mathematicians unable to cope with the complexities of an ordered, but exceedingly complex, universe.

As is usual in basic controversies, both sides claim the mantle of TRUTH.

The attitude to which we shall adhere officially, (quite apart from any psychological yearnings) is that it is presumptuous of anyone, no matter how learned, to take any definitive stand on a question of this type. On the basis of a few thousand years of scientific effort, in the presence of enigmas, riddles, and paradoxes of all varieties, we can hardly aspire to any ironclad interpretation.

Instead, we shall regard all that has been presented so far in this volume,

131

and all that will follow, as merely a set of conceptual and mathematical tricks, a bit of scientific legerdemain, which, for reasons unknown, but for which we are profoundly grateful, prove to be startlingly more successful in many applications than anyone who examines the basic assumptions would have any right to expect.

All in all, it would seem a bit unfair to sell the scientific birthright of our descendants by renouncing *forever* the possibility of precise prediction of physical phenomena. It is merely necessary that we act upon this premise at the present time. Sufficient unto the day thereof.—

9.4 Probability

There are many competing philosophical and mathematical models of uncertainty, providing room for unlimited debate concerning their individual meanings, merits, and interrelations. The concept which superficially unites them is that of *probability*. This concept enables us, at some expense as we shall see, to introduce quantitative reasoning into qualitative domains, a point which was first emphasized by Laplace in his researches in this field.

Here, we shall discuss two ways of introducing this measure of uncertainty into physical problems involving unknown aspects. Both are equally intuitive, plausible, and reasonable—and both are subject to grave criticism.

Although it may be rather disturbing to the reader, perhaps even productive of certain traumatic effects, to have pointed out constantly these cracks in the beams of our mathematical and physical edifice, and to have his attention continually directed to unsightly buttresses here and there, we feel that it is absolutely essential. Without an understanding of the flaws in the present structure, and equally without an understanding of ultimate hopes and aims, there is certainly little motivation for further digging and continued construction. There are those who are content to polish to a fine shine that which already exists, an important and useful occupation, but the true creative spirit wants to go on and construct something even more beautiful—and what can be more beautiful than a deeper understanding of the world which surrounds us.

Finally, apart from these esthetic considerations, we are forced by practical necessity to devise more efficient mathematical models. The problems which press upon us and nag at our conscience do not permit laziness or encourage smugness.

9.5 Enumeration of Equally Likely Possibilities

Let us now turn to the mathematical formulation of a theory of chadce events, and begin by considering some of the fundamental ideas.

To introduce the notion of *equally likely*, let us return to the coin-tossing operation described in the preceding section, and consider the concept of a *fair* coin. By this, we mean that we have no reason to suspect that the coin will land heads rather than tails. In this situation, we say that the *probability* of the occurrence of a head is one half, and similarly that the probability of the occurrence of a tail is one half.

In the same vein, given a die with the numbers 1, 2, 3, 4, 5, 6 on the six sides, we say that the *probability* of any particular number arising on a throw is one-sixth, provided that we have no particular reason for supposing that one number will appear rather than another.

These two examples show us how to construct a general method for assigning probabilities to the outcomes of a transformation of a physical system. Given a system S in state i, we examine the number of possible states into which the system can be transformed as a result of a particular action. Let these states be designated by the integers $1, 2, \ldots, N$, and let m_j denote the number of ways in which the system can find itself in state j, as a result of the designated cause.

Then, under our assumption of equal likelihood, we define the probability that the system will be in state j as a result of the action to be the quantity

$$(9.1) \qquad p_j = m_j \bigg/ \left(\sum_{j=1}^{N} m_j \right).$$

Observe that these quantities $p_1, p_2, \ldots, p_N$ satisfy the two conditions

$$(9.2) \quad (a) \quad p_j \geq 0,$$

$$(b) \quad \sum_{j=1}^{N} p_j = 1.$$

Generally, if we start from a state i, these quantities m_j and p_j will depend upon i. In many cases, we start from some uniform state, as in tossing a coin, so that this actual dependence need not be made explicit.

As an example of the calculation of these probabilities, given two honest* dice, we can assert that the probability of the 1–1 combination, 'snake-eyes," is $\frac{1}{36}$, while the probability of rolling a seven or an eleven on the first roll (an automatic win in craps) is $\frac{1}{6} + \frac{1}{18} = \frac{2}{9}$.

With this foundation, the usual rules are applied to determine the probabilities of compound events and so on. Furthermore, it is an easy step to the introduction of probabilities for systems which can exist in a denumerable set of states.

For the case where there are a continuum of states, certain difficulties arise which are best overcome by the use of measure theory. Since we are primarily concerned with numerical solutions, forcing us essentially to finite ranges and discrete distributions, we can, if we wish, bypass the whole subject.

---

* In 9.6 we face the difficulty of deciding when dice are "honest" or not.

It will, however, simplify some of our notation if we allow a continuum of states, or at least use Riemann-Stieltjes notation.

To extend the notion of probability to situations in which the state of a system may be any value on an interval, with the corresponding multi-dimensional vector version, we proceed as follows. Let $g(s)$ be a non-negative function defined over the interval of interest, $[a, b]$, and satisfying the two conditions

(9.3)
$$g(s) \geq 0,$$

$$\int_a^b g(s) \, ds = 1.$$

Then, we define the probability that a point t in $[a, b]$ will lie in the interval $[c, d]$ to be the integral

(9.4)
$$\int_c^d g(s) \, ds.$$

Observe that one gains the freedom of a continuum of states at the expense of giving up the probability of being in a particular state. Analogously, using Lebesgue integration, we can define the probability of a point belonging to a prescribed set in the interval $[a, b]$. Finally, using the Riemann-Stieltjes or Lebesgue-Stieltjes integral, we can combine both discrete and continuous formulations.

In what follows, we shall occasionally use the Riemann-Stieltjes integral. The reader who is unacquainted with this very useful concept should by all means become familiar with it. While doing so, he can regard all expressions of the form $dG(s)$ to be equivalent to $g(s) \, ds$, without doing violence to any of the results that follow.

The theory of probability can, and has been, introduced in quite abstract and abstruse terms. There are some advantages in doing this, but none from our point of view, since we are not interested in the theory per se, but rather in using it as a tool in the study of control and decision processes. Consequently, since we do not wish to pile the inherent complications of continuous stochastic processes upon the sufficiently complicated study of control processes, we shall also simplify the probabilistic aspects as much as possible. There is little difficulty in adding this sophistication once the new ideas are firmly grounded. To quote Hurwitz's dictum, via Polya, "It is always more difficult to particularize than to generalize."

9.6 The Frequency Approach

The question as to how one determines whether or not two outcomes are indeed equally likely must now be faced. We know from the discussion of the coin-tossing operation given above that even in this simple case we possess no combination of a priori physical measuring techniques and mathematical theory which we can use to discover whether or not the coin is biased or unbiased.

It follows that we must resort to an experimental approach, a very simple example of a Monte Carlo technique, in order to answer this question.

Let us note parenthetically that stating that a coin is fair does not, of course, mean it must come up heads on the next toss if tails have appeared on the previous toss, nor that heads is more likely on the next toss if tails have appeared on the previous ten tosses. It is remarkable to note how many betting methods employed by gambling addicts presume that a coin has a memory!

What we do mean by the concept of a fair coin is that if the coin is tossed a large number of times, then the ratio of the number of heads to the number of tails will be approximately one. Although this is intuitively sensible, the question remains as to how to make precise this new concept, based upon testing.

To being with, it is clear that we can never settle the problem experimentally in any definitive fashion. At the end of any number of trials, ten, a thousand, or a million or more, the question will still remain as to whether we have an unbiased coin with a likely sequence of tosses, or a biased coin with an unlikely sample. However, terms such as "likely" and "unlikely" can themselves be evaluated only in terms of probability theory.

As is usual in the study of the foundations of any subject, we find ourselves in the rather embarrassing position of attempting to explain simple concepts in terms of far more sophisticated ideas.

Furthermore, even the concept of testing has its limitations. For, as far as the experimental approach is concerned, we may find ourselves in a situation where even one test disturbs the system in an irreversible fashion, or where there is a chance of any single test changing the properties of the object being tested.

The only way to cut this Gordian knot is to employ the axiomatic approach, quite baldly admitting the existence of undefinable, but quite intuitive, concepts. There now remains the purely subjective problem of accepting or rejecting any particular set of axioms.

Avoiding all of this, however, consider the system mentioned above and suppose that it changes state at the time periods $0, 1, 2, \ldots$, measured in some unit of time. Observation of the system over some long time period, T, should yield experimental evidence which can be used to define what we mean by the expression "the probability that the system is in state i at time t." Furthermore, it should enable us to deduce the probability that the system will pass from state i to state j at any particular time.

The calculations are quite simple, and follow obvious lines. Over a long time period, where T tests are made, suppose that we observe the system in state i a total of T_i times. Then we can take the ratio T_i/T as an estimate for the probability that the system is in state i at any particular time.

Presumably, in the proper metric, the ratio T_i/T should converge as $T \to \infty$ to a quantity which we can define as the probability that the system is in state i at any particular time that we observe it. Counting the number of times, T_{ij}, that the system passes from state i to state j, again presumably the limit of T_{ij}/T as $T \to \infty$ should converge to the probability that the system passes from state i to state j at any time.

All of this is necessary, not so much for our immediate work, as to explain the motivation for the third part of the book—the theory of adaptive control processes. Only if we appreciate the *arbitrariness* inherent in some of the probabilistic techniques that we constantly employ, will we be properly receptive, or at least not too unfriendly, to the *arbitrary* methods that we introduce in treating adaptive processes.

9.7 Ergodic Theory

A natural question that now arises is that of the consistency of these two possible definitions of probability. This is a famous problem in both the physical and mathematical worlds, to say nothing of the philosophical and psychological domains.

It is intuitively seen that the average taken over phase space, obtained by examining the behavior at any particular time of a number of similar systems, should be equivalent to the average taken over time, obtained by examining a particular system over all time.

Nonetheless, this result has only been rigorously established for a small class of mathematical models of physical systems, and a full understanding of this equivalence remains one of the outstanding mathematical puzzles.

This equivalence is not only extremely useful for analytic purposes in giving the mathematician a choice of his field of investigation, but essential for experimental purposes. One approach, say the average over phase space, may be feasible in some situations, while the other approach, average over time, may be the only feasible approach in other cases.

9.8 Random Variables

We are now in a position to introduce the concept of a *random variable*. Let us begin with the idea of a discrete random variable. As mentioned above, if we are interested solely in computational solutions of control problems, we can restrict our attention solely to these simpler formulations. Furthermore, this combination of a discrete process over time and discrete random variables keeps the discussion on a satisfactory conceptual level. As we shall see, however, we can allow continuous random variables as long as we retain discrete processes over time. The combination of a continuous process over time and random variables, discrete or continuous, introduces complexities into which we shall not enter here, since they are in many ways extraneous to our principal goal.

We say that a quantity z is a *discrete random variable* if it assumes one of a fixed set of values $z_1, z_2, \ldots, z_M$ with the respective probabilities $p_1, p_2, \ldots, p_M$. This set of non-negative numbers, $[p_1, p_2, \ldots, p_M]$, satisfying the condition $\sum_{i=1}^{M} p_i = 1$, is called the *probability distribution* for z.

Observe that there is a considerable difference between the way in which the word "random" is used in ordinary speech and the mathematical usage, given above. For this reason, we feel that it is preferable to use the term *stochastic variable*. This adjective "stochastic" is derived from a Greek word meaning to guess or conjecture. It has the advantage of not occurring in ordinary usage, and thus, of having no intuitive associations. The scientist may scorn these semantic sensibilities, but this is a mistaken snobbery. There is no reason why a scientific term should not say what it means and mean what it says.

If two random variables, z and w, are such that a knowledge of the value of one in no way influences the distribution of values of the other, they are said to be *independent*; if not, they are called *dependent*.

Having defined stochastic quantities, we now introduce stochastic vectors as vectors whose components are stochastic variables. When we speak of a system in a random or stochastic state, we mean a system whose state is specified by a stochastic vector.

As some simple examples of stochastic variables, let us consider the variable z which can assume only two values, 1 and 0 and suppose that it assumes the first with probability p and the second with probability $1 - p$. This quantity z is a stochastic variable associated with the tossing of a biased coin where the appearance of a tail is signified by the value 1 and the appearance of a head by the value 0. If $p = \frac{1}{2}$, then, of course, we have a fair coin.

Similarly, associated with the usual die is the stochastic variable which takes the values 1, 2, 3, 4, 5, 6 with equal probability, one-sixth.

9.9 Continuous Stochastic Variable

If z is a quantity which can assume one of a continuum of values, we can introduce its probability distribution either in terms of the Riemann or Lebesgue integrable density function $g(s)$, as above, or in terms of a Stieltjes integral. As pointed out above, the Stieltjes integral enables us to handle both cases, and combinations, with the same notation. Sometimes, we shall employ this simplified notation and sometimes not. What we continually wish to emphasize is that it is essential to separate the well-known difficulties intrinsic to continuous stochastic processes from the new

137

difficulties introduced by stochastic control processes. Once these new ideas are understood, there is little difficulty in combining them with the most sophisticated of the classical ideas.

9.10 Generation of Random Variables

An interesting point to consider is that it is impossible to generate random numbers by means of mathematical and physical devices. We can by various means, some quite ingenious, generate sequences which behave in many ways like sequences of random numbers, but these are not genuine random sequences.

The subject is one of great interest and importance, and the interested reader may consult the references given at the end of the chapter. One amusing fact is that various results in the theory of numbers, seemingly far removed from reality, are often used to generate these pseudo-random sequences, which are then used, in Monte Carlo techniques, to solve problems in engineering and physics.

9.11 Stochastic Process

A process which, starting with one stochastic sequence, generates another stochastic sequence will be called a *stochastic process*. One of the simplest, and to us, for our present purposes, the most interesting, ways of generating a random sequence, $\{z_n\}$, is by means of a recurrence relation of the form

$$(9.5) \qquad z_{n+1} = g(z_n, r_n), \quad z_0 = c,$$

where the quantities $\{r_n\}$ are independent random variables drawn from a common distribution, or perhaps a distribution which itself depends upon time.

Given the properties of the r_n, the problem is that of determining the properties of the z_n. Often the z_n will have quite simple asymptotic properties as $n \to \infty$, regardless of the complexity of the r_n. This fact explains much of the success of theoretical physics.

9.12 Linear Stochastic Sequences

A particularly important type of stochastic process, arising in many different ways, is one which is described by a linear recurrence relation

$$(9.6) \qquad z_{n+1} = A_n z_n + r_n.$$

Reverting to vector-matrix notation, the quantities z_n and r_n are to be taken as N-dimensional vectors, and the A_n as $N \times N$ matrices.

Equations of this type enter conceptually in two diverse, but equally important, ways. In the first place, a deterministic system, ruled by the equation $x_{n+1} = Ax_n$ may be subjected to variable external influences.

This situation gives rise to an equation of the form

$$(9.7) \qquad z_{n+1} = Az_n + r_n, \quad z_0 = c,$$

where each r_n is a stochastic vector.

Alternatively, we may consider there to be no external forces, but allow the internal mechanism to vary according to unknown causes which we assume to be stochastic. A simple mathematical model of this yields the equation

$$(9.8) \qquad z_{n+1} = A_n z_n, \quad z_0 = c,$$

where the A_n are stochastic matrices, independent or not.

Problems of this latter type are very much more difficult to treat than those described by the equation in (9.7), because of the non-commutativity of matrix multiplication. Practically nothing has been done in connection with the study of the asymptotic behavior of the vector z_n determined by (9.8). Most of the work in this field has been devoted to the study of the sequence determined by (9.7). The simplest case of this is where $z_n = \sum\limits_{k=1}^{n} r_k$. This corresponds to an additive theory of random effects, whereas (9.8) corresponds to a multiplicative theory. The latter is a more realistic approach, whereas the former may be considered to be a perturbation theory.

9.13 Causality and the Markovian Property

The difference equations, and their limiting forms, differential equations, discussed in Chapter I gave rise, as a consequence of the associated existence and uniqueness theorems, to a principle of causality. Given the recurrence relation

$$(9.9) \qquad x_{n+1} = g(x_n), \quad x_0 = c,$$

or the differential equation

$$(9.10) \qquad \frac{dx}{dt} = g(x), \quad x(0) = c,$$

we realize that a knowledge of the state vector at any time t_0 rigidly determines the state vector at any time $t > t_0$. It follows from this that once the state at time t_0 is known, we can throw away all information concerning the states of the system at times prior to t_0 without any loss of information.

In a stochastic process, we know that the future is not precisely determined by the present. However, we can obtain a completely analogous statement if we restrict our attention to the appropriate stochastic processes, and redefine the term *state*.

139

To begin with, let us consider recurrence relations of the form appearing in (9.5) in which the $\{r_n\}$ constitute a set of independent random variables. Secondly, in place of regarding the vector given by (9.5), a stochastic vector, as representing the state of the system, let us consider the probability distribution for the components of the vector as the true state of the system.

With these conventions, we have a precise analogue of the causality emphasized in Chapter II. Given a recurrence relation of the type appearing in (11.1), where the r_n are independent stochastic variables, it is clear that the probability distribution at any time $t > t_0$ is uniquely determined by the probability distribution at time t_0.

Furthermore, given the probability distribution at t_0, we can throw away all information concerning these generalized states at time $t < t_0$ without loss of information. This generalized causality relation is usually called a *Markovian property* and processes with this property are called Markovian processes or *Markoff processes*. As usual in mathematical attribution, this is somewhat of a misnomer since Poincaré first indicated the importance of processes of this nature.

Despite the underlying uncertainty inherent in stochastic processes, by means of our mathematical artifices and artful relabelling we have restored the hoped-for causality. In the next chapter, using these new labels, we shall show how the formalism of dynamic programming carries over in a routine way to the study of stochastic control processes. Before that, we wish to discuss various other interesting aspects of stochastic processes, devoid of optimization.

9.14 Chapman-Kolmogoroff Equations

Let us now derive some analytic consequences of generalized causality, completely analogous to the results presented in 2.4.

To do this, introduce the following function of four variables:

(9.11) $p(i, m; j, n) =$ the probability that the system S is in state i at time m, given that it is in state j at time $n < m$.

To obtain the basic equation governing this function, we use the same type of diagram as in Fig. 10, albeit in different notation, since we are talking at the moment in discrete terms and we are operating in a different space.

Standard results concerning the composition of probabilities yield the relation

$$(9.12) \quad p(i, m_1 + m_2; j, n) = \sum_{k=1}^{M} p(i, m_1 + m_2; k, m_2) p(k, m_2; j, n),$$

for $m_1 > 0$ and $m_2 > n$.

This rather complicated analytic result is the mathematical translation of the following obvious statement: "The probability of being in state i

at time $m_1 + m_2$, having started in state j at time n, is derived by adding up the probabilities of going from i to k at time m_2, and then going from k to j in time m_1, for $k = 1, 2, \ldots, M$."

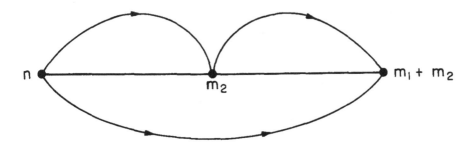

Figure 19

The point of all this is that a suitable definition of "state" and generalized causality enable us to employ the same fundamental semi-group techniques employed in the deterministic case. Abstractly, the equation above is precisely the equation derived in 2.4. It is essential to point out repeatedly basic conceptual identities of this type, since otherwise, superficial analytic differences can easily obscure the fundamental features.

9.15 The Forward Equations

Suppose that we suppress the dependence upon j and n by keeping a fixed origin in time and a fixed initial state. Taking $m_1 = 1$, we obtain the equation

$$(9.13) \qquad q(i, m_2 + 1) = \sum_{k=1}^{M} p(i, m_2 + 1; k, m_2)q(k, m_2),$$

where we have simplified the notation by writing

$$(9.14) \qquad q(i, m) = p(i, m; j, n).$$

The coefficients $p(i, m_2 + 1; j, m_2)$ are called *transition probabilities*. In many important situations, they are independent of time, i.e. of m_2, so that we may write the recurrence relations in the form

$$(9.15) \qquad q(i, m_2 + 1) = \sum_{k=1}^{M} p_{ik}q(k, m_2),$$

or, finally reducing i to a subscript, and using the variable n in place of m_2, in the standard form for a finite-dimensional Markoff process,

$$(9.16) \quad x_i(n + 1) = \sum_{k=1}^{M} p_{ik}x_k(n), \quad n = 0, 1, \ldots, i = 1, 2, \ldots, M.$$

141

The asymptotic behavior of the state vector $x(n)$ depends upon the algebraic properties of the matrix $P = (p_{ik})$ in a rather complex way. Superficially, the dependence seems simple, since we can write $x(n) = P^n x(0)$. These questions will be discussed in further detail in Chapter XI.

Similarly, keeping both the terminal state and terminal time fixed, we can define the function

$$(9.17) \qquad r(j, n) = p(i, m; j, n)$$

and obtain equations analogous to those given in (9.13). These are called the *backwards equations*.

In some investigations, it is more convenient to use one set, in some investigations the other. Since these equations are dual to each other in the usual analytic sense of duality, it is often most convenient to use them in combination.

9.16 Diffusion Equations

One of the most interesting and useful equations of mathematical physics is the equation of heat conduction and diffusion theory, of which the simplest version is the equation

$$(9.18) \qquad u_t = u_{xx}.$$

What is most remarkable, and the basis for many theoretical investigations and computational techniques based upon Monte Carlo principles, is that this deterministic equation is the limiting form of equations associated with stochastic processes.

In 2.5 it was shown that a limiting form of the basic functional equation derived from semi-group principles, applied to the deterministic processes discussed there, was a linear partial differential equation. Similarly, by introducing a continuum of states and continuous time, and then using a careful limiting procedure, the discrete Chapman-Kolmogoroff equations of 9.15 will yield the heat equation written above, and, of course, more general parabolic equations in more general geometric setting.

The importance of this approach from the conceptual point of view is great. Many important properties of the solutions of parabolic equations can be understood immediately and can, with the use of modern techniques of probability theory, actually be derived rigorously from a consideration of the original stochastic process. Furthermore, as mentioned above, this interconnection offers an entirely new computational approach to the solution of equations of this nature.

9.17 Expected Values

Starting with our usual recurrence relation, a knowledge of the distribution for the stochastic variables appearing in this equation, and given the distribution of values for the initial state, our paramount concern is that of obtaining the distribution of outcomes at the end of the n-th stage of the process. The Chapman-Kolmogoroff equations enable us to settle this question, at least analytically.

A probability distribution, however, is a rather complicated bit of information to cart about. Consequently, on both analytic and computational grounds, we quite often agree to settle for a simpler measure of the actual state of the system.

The simplest, (and most useful, as a consequence of analytic properties mentioned below and theoretical asymptotic results we shall not mention at all), of these substitute measures is the *expected value* of z, written $E(z)$ or Exp (z). It is defined by the relation

$$(9.19) \qquad E(z) = \sum_{i=1}^{N} p_i z_i,$$

if the distribution of values is discrete, and by the expression

$$(9.20) \qquad E(z) = \int_{-\infty}^{\infty} z g(z) \, dz,$$

if the distribution is continuous with a density function, $g(z)$. Generally, it has the form of a Stieltjes integral

$$(9.21) \qquad E(z) = \int_{-\infty}^{\infty} z \, dG(z).$$

In other words, $E(z)$ is the weighted average of z. It is also called the first *moment*, by analogy with the mechanical situation where weights p_i are located at distances z_i from the axis.

Of great interest also is the second moment

$$(9.22) \qquad E(z^2) = \sum_{i=1}^{N} p_i z_i^2, \quad \text{or} \quad \int_{-\infty}^{\infty} z^2 \, dG(z),$$

the weighted average of z^2, since with $E(z)$ and $E(z^2)$, we can evaluate

$$(9.23) \qquad E[\{z - E(z)\}^2] = E(z^2) - 2E(z)^2 + E(z)^2$$
$$= E(z^2) - E(z)^2.$$

This quantity, the *variance*, provides a great deal of information concerning the distribution of z about its expected value.

The principal reason why the expected value is used so frequently lies in its *linearity*. Whether z and w are dependent or independent, we have the fundamental relation

$$(9.24) \qquad E(z + w) = E(z) + E(w).$$

This is extremely important in connection with our subsequent study of stochastic control processes where highly dependent sequences of random sequences are generated.

One of the great advantages, however, of the functional equation technique that we shall employ is that we can emancipate ourselves from over-dependence upon linear relations and linear criteria. We can use other measures apart from expected value without affecting the computational feasibility of a solution.

Finally, let us note that if z and w are independent, then

$$(9.25) \qquad \mathrm{E}(e^{i(z+w)t}) = \mathrm{E}(e^{izt})\mathrm{E}(e^{iwt}),$$

a relation of which (9.24) is a limiting case as $t \to 0$. A useful rule-of-thumb principle in dealing with independent random variables is that whatever can be obtained using the linearity relation for expected values can also be derived for the more informative quantity $\mathrm{E}(e^{izt})$.

9.18 Functional Equations

Let us now obtain the analogues of the equations derived in 2.4 in connection with deterministic processes. Once again, we employ the basic idea of functional dependence.

Consider the recurrence relation

$$(9.26) \qquad z_{n+1} = g(z_n, r_n), \quad z_0 = c,$$

where the $\{r_n\}$ constitute a set of independent random variables. We observe that $\mathrm{Exp}\,[h(z_n)]$, where $h(y)$ is a given function of the vector y (scalar or vector), depends only upon $z_0 = c$, the initial vector, and n, the time variable. Let us then write

$$(9.27) \qquad f_n(c) = \mathrm{Exp}\,[h(z_n)],$$

for all c, and $n = 1, 2, \ldots$. The expected value is taken over the distributions determining the stochastic vectors $r_1, r_2, \ldots, r_n$. Henceforth, we shall write Exp for the expected value in place of E in order to indicate quite clearly that we are taking an expected value.

We then have the fundamental recurrence relation

$$(9.28) \qquad f_n(c) = \mathrm{Exp}_{r_1}\,[f_{n-1}\{g(c, r_1)\}],$$

where now, as the notation indicates, the expected value is taken only over r_1. Using Stieltjes integral notation, this takes the form

$$(9.29) \qquad f_n(c) = \int_{-\infty}^{\infty} f_{n-1}\{g(c, r_1)\}\, dG(r_1),$$

where $dG(r_1)$ is the distribution function associated with the random variable r_1. If r_1 assumes only two values, say ± 1, with equal probability, this reduces to

$$(9.30) \qquad f_n(c) = \tfrac{1}{2}[f_{n-1}\{g(c, 1)\} + f_{n-1}\{g(c, -1)\}].$$

144

The stochastic equation of (9.26) thus gives rise to an extension of the concept of iteration. The relation in (9.28) is an analogue of equation (2.3).

9.19 An Application

Using relations of this type, we have a powerful way of using digital computers to study the behavior of nonlinear systems subject to stochastic effects.

For example, to discuss the behavior of a multivibrator, governed by the Van der Pol equation with a stochastic forcing function

$$(9.31) \qquad u'' + \lambda(u^2 - 1)u' + u = r(t),$$

we use the discrete approximating version

$$(9.32) \qquad u_{n+1} - u_n = v_n\Delta, \quad u_0 = c_1,$$
$$v_{n+1} - v_n = [r_n - u_n - \lambda(u_n^2 - 1)v_n]\Delta, \quad v_0 = c_2,$$

and proceed as above.

Let

$$(9.33) \qquad f_n(c_1, c_2) = \text{Exp}\,\{g(u_n, v_n)\},$$

for $n = 1, 2, \ldots, -\infty < c_1, c_2 < \infty$. Then, using the foregoing arguments, we have the recurrence relation

$$(9.34) \quad f_n(c_1, c_2) = \underset{r_1}{\text{Exp}}\,[f_{n-1}(c_1 + c_2\Delta, c_2 + \{r_1 - c_1 - \lambda(c_1^2 - 1)c_2\}\Delta)],$$

for $n \geq 1$. If $r_1 = \pm 1$ with equal probability, then this reads

$$(9.35) \quad f_n(c_1, c_2) = \tfrac{1}{2}[f_{n-1}(c_1 + c_2\Delta, c_2 + \{1 - c_1 - \lambda(c_1^2 - 1)c_2\}\Delta)$$
$$+ f_{n-1}(c_1 + c_2\Delta, c_2 + \{-1 - c_1 - \lambda(c_1^2 - 1)c_2\}\Delta)],$$

for $n \geq 1$, with $f_0(c_1, c_2)$ determined by

$$(9.36) \qquad f_0(c_1, c_2) = g(c_1, c_2).$$

In this form, we have a recurrence relation which can be used in connection with a digital computer to determine a number of important properties of the stochastic process. A reference to some particular computations will be found at the end of the chapter.

9.20 Expected Range and Altitude

In Chapter II we showed how we could use the functional dependence upon initial values to study problems of maximum range and altitude. In exactly the same way, if the governing equations contain stochastic elements, we can compute the expected maximum altitude and expected range. We shall consider an implicit stochastic variational problem in the next chapter to illustrate this technique.

Bibliography and Comments

9.1 For a detailed discussion of many of the matters that are treated in summary fashion here, see

P. S. Laplace, *A Philosophical Essay on Probabilities*, Dover Publications, New York, 1951.

I. J. Good, *Probability and the Weighing of Evidence*, C. Griffin and Co., London, 1950.

I. J. Good, "Kinds of probability," *Science*, vol. 129, 1959, pp. 443–447.

R. A. Fisher, "Mathematical probability in the Natural Sciences," *Technometrics*, vol. 1, 1959, pp. 21–30,

N. T. Gridgeman, "The lady testing tea and allied topics," *J. Amer. Stat. Assoc.*, vol. 54, 1959, pp. 776–783.

O. Ore, "Pascal and the invention of Probability Theory," *Amer. Math. Monthly*, vol. 67, 1960, pp. 409–418.

and the book

H. L. Jeffreys, *Statistical Inference*, Cambridge Univ. Press, 1957.

For a lively discussion of a supposed difference between the social and unsocial sciences (or perhaps even antisocial), see

M. Scriven, "Explanation and prediction in evolutionary thought," *Science*, vol. 130, 1959, pp. 477–481.

We don't really expect any reader to absorb the foundations of the theory of probability from these brief comments and definitions. We do, however, wish to review and make precise the assumptions under which we are operating so that our further discussion will be more understandable and reasonable.

9.2 Concerning the manifold ways in which the question "What is the probability of this event?" can be interpreted, it is perhaps appropriate to quote from a letter from Newton to Pepys in response to a problem posed by the latter:

". . . I do not see that the words of the question, as 'tis set down in your letter, will admit; but this being no mathematical question, but a question what is the true mathematical question, it belongs not to me to determine it."

This quotation, the inclusion of which I owe to the kindness of J. W. T. Youngs and M. Zorn, it taken from p. 253 of vol. IV of *Diary and Correspondence of Samuel Pepys . . . Life and Notes by Lord Braybrooke*, The C. T. Brainard Publishing Co., Boston-New York.

It is perhaps also appropriate to add that we do not at all agree with this point of view. We feel that perhaps the most important task and responsibility of the mathematician is that of posing the "correct" problem. This attitude on Newton's part is consistent with his famous dictum "Non fingo hypotheses."

For a further discussion of stability, see

R. Bellman, *Stability Theory of Differential Equations*, McGraw-Hill Book Co., Inc., New York, 1954.

9.3 See the interesting book

D. Bohm, *Causality and Chance in Modern Physics*, Van Nostrand, 1957.

9.4 For an introduction to the theory of probability and its analytic aspects, see

J. V. Uspensky, *Introduction to Mathematical Probability*, McGraw-Hill Book Co., Inc., New York, 1937.

W. Feller, *Probability Theory and Its Applications*, John Wiley and Sons, New York, 1957.

9.7 For the mathematical treatment of ergodic theory, see the monograph
E. Hopf, *Ergodentheorie*, Ergebnisse der Math., vol. 2, J. Springer, Berlin, 1937,

and the following two works:

N. Dunford and J. T. Schwartz, *Linear Operators, Part I, General Theory*, Interscience, 1957,

P. R. Halmos, *Lectures on Ergodic Theory*, Math. Soc. Japan, Tokyo, 1956,

where references to the fundamental results of G. D. Birkhoff and J. von Neumann may be found, together with numerous extensions.

9.10 For the theory and application of Monte Carlo techniques, see

H. A. Meyer, editor, *Symposium on Monte Carlo Methods*, J. Wiley and Sons, New York, 1956.

9.12 For a treatment of linear systems with random coefficients, see

L. Zadeh, "Correlation functions and power spectra in variable networks," *Proc. IRE*, vol. 38, 1950, November.

A. Rosenbloom, *Analysis of Randomly Time Varying Linear Systems*, Ph.D. Thesis, U.C.L.A., 1954.

J. C. Samuels and A. C. Eringen, *On Stochastic Linear Systems*, Technical Report No. 11, Purdue Univ., 1957,

where many other references may be found. See also

R. Bellman, *Matrix Analysis*, McGraw-Hill Book Co., Inc., New York, 1960, for a large number of additional references.

9.14–9.16 See

A. Khintchine, *Asymptotische Gesetze der Wahrscheinlichkeitsrechnung*, Chelsea, 1948.

M. Kac, "On some connections between probability theory and differential and integral equations," *Second Berkeley Symposium on Math. Stat. and Probability*, Univ. of Calif. Press, 1951.

9.19 See the paper by

R. Bellman, P. Brock and M. Mizuki, "Computational determination of the nature of the solutions of nonlinear systems with stochastic inputs," AIEE Special Publication T-101, *Proc. Computers in Control Systems Conference*, Atlantic City, October 1957, pp. 109–110.

9.20 See

R. Bellman, "Functional equations and maximum range," *Q. Appl. Math.*, vol. 17, 1959, pp. 231–236.

R. Bellman and J. Richardson, "On the application of dynamic programming to a class of implicit variational problems," *Q. Appl. Math.*, to appear.

There is a famous couplet of Pope's which reads:

> Nature, and Nature's laws, lay hid in night;
> God said, Let Newton be! and all was light.

To this, J. C. Squire added:

> It did not last: the Devil, howling Ho!
> Let Einstein be! restored the status quo.

(We take this second quotation from the heading to Chapter VIII of the delightful book,

H. Jeffreys, *Scientific Inference*, Cambridge, 1957.)

To these four lines we might add:

> Then God and Devil, bo(h)red, revealed the quantum,
> To haunt excited physicists ad infinitum.

(I wish to thank H. S. Bailey, Jr., for the polished form of this couplet.)

Finally, while in a poetic mood, let us quote the following poem by M. G. Kendall, which appeared in the *American Statistician*, vol. 13, 1959, pp. 23–24. It expresses quite succinctly some of the ideas over which we have worried in the foregoing pages.

HIAWATHA DESIGNS AN EXPERIMENT
Maurice G. Kendall

1. Hiawatha, mighty hunter
   He could shoot ten arrows upwards
   Shoot them with such strength and swiftness
   That the last had left the bowstring
   Ere the first to earth descended.
   This was commonly regarded
   As a feat of skill and cunning.

2. One or two sarcastic spirits
   Pointed out to him, however,
   That it might be much more useful
   If he sometimes hit the target.
   Why not shoot a little straighter
   And employ a smaller sample?

3. Hiawatha, who at college
   Majored in applied statistics
   Consequently felt entitled
   To instruct his fellow men on
   Any subject whatsoever,
   Waxed exceedingly indignant

Talked about the law of error,
Talked about truncated normals,
Talked of loss of information,
Talked about his lack of bias
Pointed out that in the long run
Independent observations
Even though they missed the target
Had an average point of impact
Very near the spot he aimed at
(With the possible exception
Of a set of measure zero).

4. This, they said, was rather doubtful.
Anyway, it didn't matter
What resulted in the long run;
Either he must hit the target
Much more often than at present
Or himself would have to pay for
All the arrows that he wasted.

5. Hiawatha, in a temper
Quoted parts of R. A. Fisher
Quoted Yates and quoted Finney
Quoted yards of Oscar Kempthorne
Quoted reams of Cox and Cochran
Quoted Anderson and Bancroft
Practically in extenso
Trying to impress upon them
That what actually mattered
Was to estimate the error.

6. One or two of them admitted
Such a thing might have its uses
Still, they said, he might do better
If he shot a little straighter.

7. Hiawatha, to convince them
Organized a shooting contest
Laid out in the proper manner
Of designs experimental
Recommended in the textbooks
(Mainly used for tasting tea, but
Sometimes used in other cases)
Randomized his shooting order
In factorial arrangements
Used in the theory of Galois
Fields of ideal polynomials
Got a nicely balanced layout
And successfully confounded
Second-order interactions.

8. All the other tribal marksmen
Ignorant, benighted creatures,
Of experimental set-ups

Spent their time of preparation
Putting in a lot of practice
Merely shooting at a target.

9. Thus it happened in the contest
That their scores were most impressive
With one solitary exception
This (I hate to have to say it)
Was the score of Hiawatha,
Who, as usual, shot his arrows
Shot them with great strength and swiftness
Managing to be unbiased
Not, however, with his salvo
Managing to hit the target.

10. There, they said to Hiawatha,
This is what we all expected.

11. Hiawatha, nothing daunted,
Called for pen and called for paper
Did analyses of variance
Finally produced the figures
Showing beyond peradventure
Everybody else was biased
And the variance components
Did not differ from each other
Or from Hiawatha's
(This last point, one should acknowledge
Might have been much more convincing
If he hadn't been compelled to
Estimate his own component
From experimental plots in
Which the values all were missing.
Still, they didn't understand it
So they couldn't raise objections
This is what so often happens
With analyses of variance).

12. All the same, his fellow tribesmen
Ignorant, benighted heathens,
Took away his bow and arrows,
Said that though my Hiawatha
Was a brilliant statistician
He was useless as a bowman,
As for variance components
Several of the more outspoken
Made primeval observations
Hurtful of the finer feelings
Even of a statistician.

13. In a corner of the forest
Dwells alone my Hiawatha
Permanently cogitating

On the normal law of error
Wondering in idle moments
Whether an increased precision
Might perhaps be rather better
Even at the risk of bias
If thereby one, now and then, could
Register upon the target.

From *The American Statistician*
Vol. 13, No. 5, 1959, pp. 23–24.

CHAPTER X

STOCHASTIC CONTROL PROCESSES

Quid facerem? neque servitio ne exire licebat
nec tam praesentis alibi cognoscere divos.
What was I to do? I could not quit my slavery
nor elsewhere find my gods so ready to aid.

Virgil, Ecologues, 40–42.

10.1 Introduction

In the preceding chapter, we discussed in some detail why we were forced
to renounce determinism in treating various physical processes. We then
turned to the concept of random variables and probability distributions,
albeit with various misgivings concerning the relevancy of this mathematical structure to actual processes.

Let us now put aside these worthy doubts and consider these stochastic
processes as deserving of investigation in their own right. As we shall see
in this and succeeding chapters, stochastic control processes yield many
fascinating fields of study with apparently unending ramifications, and
certainly no visible boundaries at the present time.

In this chapter, we wish to consider certain fundamental questions in the
mathematical theory of stochastic control processes. Our aim is to show that
the functional equation technique of dynamic programming is applicable
to these arcane areas, and, indeed, is applicable in a routine way to problems
of formulation and problems of computation.

We have avoided completely the subject of stochastic control processes
of continuous type since these combine the subtleties and complexities of
the calculus of variations and the theory of continuous stochastic processes.
This is a vital and interesting field of research in which practically nothing
has been done. As stated before, our aim is to keep separate the complexities
inherent to control processes from those ever present in the study of continuous stochastic processes.

10.2 Discrete Stochastic Multistage Decision Processes

In order to make the analogy with the previous discussion of deterministic
control processes as apparent as possible, let us parallel the route followed
in Chapter III.

To begin with, we must define what we mean by a *discrete stochastic process*. The adjective "discrete," as before, signifies a process in which the transformations occur at a finite, or, at worst, enumerable set of times. By "stochastic," we again mean that a decision, q, determines not a unique outcome, but a set of possible outcomes, $\{T(p, q, r)\}$, where r is a stochastic vector with a given distribution function.

Let us now insert the multistage aspects. At an initial time, and an initial state, which we call p_1, an initial decision q_1 is made. The result is a stochastic state, $p_2 = T(p_1, q_1, r_1)$.

It is important to emphasize the fact that although only the distribution of outcomes, p_2, is known *before* the choice of q_1, *after* this decision has been made we are allowed to observe the new state of the system. We underline this point since there are a number of quite interesting and significant stochastic processes in which the state of the system is only imperfectly known at each stage of the process. Some processes of this nature will be treated subsequently. In studying these, we are forced to describe the state of the system in a much more complicated fashion. Here we consider only the simpler process in which the state of the system is completely known before each decision.

It is important to note that there is no single stochastic analogue of a deterministic control process, but rather an entire spectrum of processes to choose from, each of which corresponds in some way, and each of which reduces to a deterministic process if uncertainty is removed.

Let us now return to our multistage stochastic control process. Starting with the new state, p_2, a decision q_2 is made, resulting in a third state, $p_3 = T(p_2, q_2, r_2)$, and so on. Each state of the system after the first is a stochastic variable, and (as is worth mentioning) so also are the control vectors, q_i, since they depend, in general, upon the current state of the system.

As before, with each N-stage decision process, we associate a scalar function, the criterion function

$$(10.1) \qquad F(p_1, p_2, \ldots, p_N; q_1, q_2, \ldots, q_N; r_1, r_2, \ldots, r_N),$$

itself a stochastic quantity, which evaluates the sequence of decisions.

10.3 The Optimization Problem

Since the scalar quantity, F, is a stochastic quantity, it is generally meaningless to talk about maximizing or minimizing this expression. Instead, we must agree to use the expected value of F as a measure of utility. The average value is to be taken over the stochastic variables, $r_1, r_2, \ldots, r_N$.

Although it is intuitively clear that optimization problems of this genre are considerably more difficult than the corresponding problems for deterministic control processes, it is worth analyzing the reasons for this added complexity in some detail.

10.4 What Constitutes an Optimal Policy?

The principal reason why the stochastic variational problem posed in the preceding paragraphs is more complex than those we have previously studied is that the very concept of a solution, an optimal policy, is now vastly more subtle.

In treating deterministic processes, a policy can be conceived of a set of N *vectors*, $[q_1, q_2, \ldots, q_N]$, determined by the initial state. This is a consequence of the uniqueness of solution of the underlying equations. Alternately, a policy can be viewed as a set of N *functions*, $q_1(p_1), q_2(p_2), \ldots,$ $q_N(p_N)$, of the current state vectors. In choosing the q_i, we have a choice between choosing them simultaneously, the first approach, or choosing them one at a time, corresponding to the second approach.

The fact that this choice of methods exists is of great importance. In many cases, the first approach is simpler as far as formulation is concerned, and leads to a quicker determination of the fundamental properties of the solution. In other cases, the second approach, based upon the policy concept, is the more natural technique to employ, and occasionally one mixes the two approaches.

The second approach, however, involving as it does functions rather than variables, necessarily requires a higher plane of mathematical sophistication.

In a stochastic control process, these two formulations (which, as previously pointed out in the deterministic case, reflect the duality of Euclidean space) lead to two quite different mathematical problems, and reflect two quite different versions of the control problem.

The problem of choosing a predetermined set of vectors $[q_1, q_2, \ldots, q_N]$ which will minimize the expected value of the criterion function is a well-defined mathematical question. We shall, however, not be concerned with this problem, interesting as it is, for two reasons. In the first place, it does not involve the feedback control concept, and in the second place, it is too difficult.

We wish to consider the multistage decision process in which the q_i are stochastic variables which depend upon the current state of the system, p_i. What is rather remarkable is that this approach which would seem to be more complicated is actually easier to utilize from the standpoint of formulation, analysis, and numerical evaluation.

This is so, despite the fact stressed above that a deterministic feedback control process requires only a point in Euclidean space, $[q_1, q_2, \ldots, q_N]$, to fix an optimal policy, whereas a stochastic feedback control process requires a point in function space, $[q_1(p_1), q_2(p_2), \ldots, q_N(p_N)]$.

What is important to emphasize is that the concept of a *policy*, which appears perhaps as an artifice in the deterministic control process, looms as a necessity in the study of stochastic control processes. The functional equation approach, an alternative in the former case, appears to be the only general approach in the latter.*

10.5 Two Particular Stochastic Control Processes

Let us now describe the type of stochastic control process we wish to consider first. We shall suppose that the state of the system at the n-th time period, x_n, is determined by means of a recurrence relation of the form

$$(10.2) \qquad x_{n+1} = g(x_n, y_n, r_n), \quad x_0 = c,$$

where the r_i constitute a set of independent stochastic vectors, and the y_i are the decision vectors, a set of quite dependent stochastic vectors.

A problem of particular interest arises when we wish to minimize the expected value of a sum of the form

$$(10.3) \qquad F_1 = h(x_1, y_1, r_1) + h(x_2, y_2, r_2) + \ldots + h(x_N, y_N, r_N).$$

We can think of this expression as measuring the deviation of the system from some desired performance, or, perhaps, the return from some continuing economic process.

A particular case, of frequent occurrence, is that where we wish merely to minimize the expected value of some function of the final state,

$$(10.4) \qquad F_2 = h(x_N),$$

terminal control.

Subsequently in 10.9 and 10.10 we shall discuss more complicated processes in which we must use implicit criteria of performance.

10.6 Functional Equations

Using the criterion function defined in (10.3), let us introduce the new function

$$(10.5) \qquad f_N(c) = \underset{\{y_i\}}{\text{Min}} \left[\underset{\{r_i\}}{\text{Exp}} F_1 \right].$$

The minimum is now taken over all allowable feedback policies $y_i = y_i(p_i)$ and the expected value over the r_i, assumed for the sake of notational convenience to possess the common distribution function $dG(r)$. For $N = 1$, we have

$$(10.6) \qquad f_1(c) = \underset{y_1}{\text{Min}} \left[\underset{r_1}{\text{Exp}} \, h(c, y_1, r_1) \right]$$

$$= \underset{y_1}{\text{Min}} \left[\int h(c, y_1, r_1) \, dG(r_1) \right],$$

---

* Sweeping statements such as this must always be viewed with a healthy skepticism. What this really means is that we don't know any better methods at the present time. It follows that this is less of a dictum than a challenge.

while for $N \geq 2$, we have the recurrence relation

$$(10.7) \quad f_N(c) = \underset{v_1}{\text{Min}} \left[\underset{r_1}{\text{Exp}} \left[h(x_1, y_1, r_1) \right] + f_{N-1}[g(c, y_1, r_1)] \right]$$

$$= \underset{v_1}{\text{Min}} \left[\int \{ h(x_1, y_1, r_1) + f_{N-1}[g(c, y_1, r_1)] \} \, dG(r_1) \right].$$

As in Chapter III, we can derive this either purely analytically, following 10.7, or obtain it as a consequence of the principle of optimality, 10.8.

10.7 Discussion

It is important to observe that precisely the same techniques used to treat deterministic control processes have been applied to stochastic control processes, and completely analogous equations have been obtained.

It follows that as different as deterministic and stochastic control processes appear to be, the functional equation technique affords a unified approach which treats both types of processes in precisely the same fashion.

Abstractly, they can be regarded as processes of the same type, with suitable definitions of the state of the system appropriate to each individual process.

10.8 Terminal Control

Similarly, referring to the criterion function of (10.4), if we set

$$(10.8) \qquad\qquad f_N(c) = \underset{\{y_i\}}{\text{Min}} \underset{\{r_i\}}{\text{Exp}} \, h(x_N),$$

we have

$$(10.9) \qquad\qquad f_1(c) = h(c),$$

$$f_N(c) = \underset{v_1}{\text{Min}} \left[\underset{r_1}{\text{Exp}} f_{N-1}\{ g(c, y_1, r_1) \} \right]$$

$$= \underset{v_1}{\text{Min}} \int f_{N-1}[g(c, y_1, r_1)] \, dG(r_1).$$

For the purposes of numerical computation, $dG(r)$ will be taken to be discrete, so that the preceding relation will have the form

$$(10.10) \qquad f_N(c) = \underset{v_1}{\text{Min}} \left[\sum_{i=1}^{M} p_i f_{N-1}\{ g(c, y_1, s_i) \} \right].$$

In many cases, the probabilities p_i themselves will depend upon the decision that is made, so that (10.10) will have the form

$$(10.11) \qquad f_N(c) = \underset{v_1}{\text{Min}} \left[\sum_{i=1}^{M} p_i(y_1) f_{N-1}\{ g(c, y_1, s_i) \} \right],$$

or the still more general form in which the p_i depend upon c and y_1.

It follows that the computational effort required to solve a problem of this nature is scarcely longer than that required for a deterministic process.

In Chapter XVIII we shall show how this formula yields much simpler analytic recurrence relations in the case where $g(c, y_1, r_1)$ is linear in its variables, and $h(c)$ is quadratic.

10.9 Implicit Criteria

In a number of problems of current interest, in place of a specific analytic criterion function, we have an implicit criterion determined by the process itself. We have met one example of this in the section devoted to the "bang-bang" control problem. Generally speaking, we wish to be able to handle problems of the following nature:

Given a recurrence relation of the type appearing in (10.2), when x_n satisfies a set of conditions $C_1, C_2, \ldots, C_p$ for the first time, we want an assigned scalar function of x_n, $h(x_n)$, to have a minimum expected value.

There are a number of quite interesting existence and uniqueness questions connected with these problems which have not been discussed at all to date.

Let us illustrate the treatment of problems of this nature by means of a two-dimensional process.

Typical of a problem of this nature is that of controlling a space ship so as to minimize the landing velocity. Here the control policy determines the time of landing.

10.10 A Two-dimensional Process with Implicit Criterion

Consider the scalar equations

$$(10.12) \quad x_1(n + 1) = x_1(n) - y_1(n) - r_1(n), \quad x_1(0) = c_1,$$
$$x_2(n + 1) = g_2[x_1(n), x_2(n), y_2(n), r_2(n)], \quad x_2(0) = c_2.$$

We wish to choose the control components $y_1(n)$ and $y_2(n)$, subject to constraints of the form

$$(10.13) \quad 0 \leq a_1 \leq y_1(n) \leq a_2, \quad 0 \leq b_1 \leq y_2(n) \leq b_2,$$

so as to minimize the expected value of $[x_2(m) - x_0]^2$ where m is the first time at which $x_1(m) \leq 0$. The $r_i(n)$ are independent random variables with given distributions.

The expected value is taken over the random variables $r_1(n)$ and $r_2(n)$. Let us suppose that the control variable $y_1(n)$ and $r_1(n)$ are constrained by a relation such as

$$(10.14) \quad y_1(n) + r_1(n) \geq a_3 > 0,$$

so that $x_1(n)$ is steadily decreasing as n increases. Consequently, there is no doubt that there is a value of m such that $x_1(m) \leq 0$ for the first time.

Since the minimum of the expected value of $[x_2(m) - x_0]^2$ depends only upon c_1 and c_2, let us write

(10.15)
$$f(c_1, c_2) = \operatorname*{Min}_{y_i} \left[\operatorname*{Exp}_{r_i} [x_2(m) - x_0]^2 \right].$$

Then

(10.16)
$$f(0, c_2) = (c_2 - x_0)^2,$$

and for $c_1 > 0$, we have

(10.17)
$$f(c_1, c_2) = \operatorname*{Min}_{y_1, y_2} \left[\operatorname*{Exp}_{r_1, r_2} f[c_1 - y_1 - r_1, g(c_1, c_2, y_2, r_2)] \right].$$

In order to obtain a computational solution of this equation, we can employ successive approximations, along lines discussed subsequently, or we can, more simply, use the fact that c_1 is monotone decreasing, albeit in a stochastic fashion.

If g_2 is linear in its arguments, this relation can be greatly simplified by making use of the quadratic nature of $f(c_1, c_2)$.

Bibliography and Discussion

10.1 We follow in this chapter the presentation in

R. Bellman, *Dynamic Programming*, Princeton University Press, Princeton, N.J., 1957, Chapter 3,

and

R. Bellman, "Dynamic programming and stochastic control processes," *Information and Control*, vol. 1, 1958, pp. 228–239.

10.9 The particular problem discussed here was first treated in

R. Bellman and J. M. Richardson, "On the application of dynamic programming to a class of implicit variational problems," *Q. Appl. Math.*, to vol. 17, 1959, pp. 231–236.

A stochastic control process of great interest in economic and industrial control processes is the "optimal inventory" process. See Chapter 5 of the book cited above, and

K. Arrow, S. Karlin, and H. Scarf, *Studies in the Mathematical Theory of Inventory and Production*, Stanford Univ. Press, Stanford, California, 1958.

It would be desirable to have available a rigorous theory of the minimization of the expected value of the functional

$$J(y) = \int_0^T g[x, y, r(t)] \, dt,$$

where x and y are related by the differential equation

$$\frac{dx}{dt} = h[x, y, r(t)], \quad x(0) = c,$$

and $r(t)$ is a random function with prescribed stochastic properties.

BIBLIOGRAPHY AND DISCUSSION

For some preliminary results, see

R. Bellman, "On the foundations of a theory of stochastic variational processes," *Proc. Tenth Symposium in Applied Math.*, Amer. Math. Soc., 1961.

For some results marking the beginning of a variational theory of stochastic processes in physics, analogous to Hamiltonian theory for classical mechanics, see

L. Onsager and S. Machlup, "Fluctuations and irreversible processes—I," *Phys. Rev.*, vol. 91, 1953, pp. 1505–1515.

R. Kikuchi, *Irreversible Cooperative Phenomena*, to appear.

R. Kikuchi, "Statistical Dynamics of Boundary Motion," Research Report No. 149, June, 1960, Hughes Research Laboratories.

For another approach to the subject of stochastic control processes, see

N. Wiener, *Nonlinear Problems in Random Theory*, John Wiley and Sons, New York, 1958.

CHAPTER XI

MARKOVIAN DECISION PROCESSES

Never look behind you. Something may be
gaining on you.

SATCHEL PAIGE

11.1 Introduction

In this chapter we wish to formulate and discuss a stochastic control
process of particularly simple analytic structure. The almost-linear aspects
of the equations describing this process permit us to deduce the nature of
the asymptotic form of the return function as the number of stages increases
without limit. This, in turn, allows us to introduce with profit the significant
concept of an optimal *average return*, an idea which enables us to employ
various simple iterative techniques to determine the optimal policy without
the intervention of functions of a large number of variables. Of these
methods of successive approximation, the most important is one due to
R. Howard.

As we shall mention again below in two contexts, the steady-state process
has close connections with linear programming formulations of these
problems. We shall indicate the application of steady-state techniques to
some other interesting processes of both deterministic and stochastic
origin. These include the optimal inventory problem, bottleneck processes
of the type that arise in input-output analysis, and feedback control
processes governed by linear equations and quadratic criteria.

11.2 Limiting Behavior of Markov Processes

Let us recall the definition of a discrete Markov process. We suppose that
we have a system S capable of existing in any of a finite number of states,
which we designate by $i = 1, 2, \ldots, N$, at any particular time, and a
transition matrix $A = (a_{ij})$, where the individual element a_{ij} is the probability that the system will be found in state i at time $t + 1$ if it was in state
j at time t. Here t assumes only the values $t = 0, 1, 2, \ldots$, and we suppose
that the matrix A is independent of time. In what follows below, we shall
consider a more general case arising from a decision process.

Introducing the functions

(11.1) $x_i(t) =$ the probability that S is in state i at time t,

160

we obtain, by way of the usual composition of probabilities, the recurrence relations

$$(11.2) \qquad x_i(t+1) = \sum_{j=1}^{N} a_{ij} x_j(t), \quad i = 1, 2, \ldots, N.$$

In addition, we have the initial conditions

$$(11.3) \qquad x_i(0) = c_i, \quad c_i \geq 0, \quad \sum_{i=1}^{N} c_i = 1.$$

In vector-matrix terms,

$$(11.4) \qquad x(t+1) = Ax(t), \quad x(0) = c.$$

Perhaps the most interesting problem associated with this process is that of determining its limiting behavior as $t \to \infty$. In general, despite the fact that we can write $x(t) = A^t c$, this is a problem of some delicacy with a proliferation of cases and conditions. If, however, the transition probabilities are all positive, a very simple and satisfactory result can be obtained. If $a_{ij} > 0$ for all i and j, then we may assert that as $t \to \infty$, we have $x(t) \to y$, where y is a characteristic vector of A with associated characteristic value one.

We have

$$(11.5) \qquad Ay = y.$$

If, as is natural, $x(0) = c$ is a probability vector, i.e.

$$(11.6) \qquad c_i \geq 0, \quad \sum_{i=1}^{N} c_i = 1,$$

then y will also be a probability vector.

It is easy to show that y is unique if all of the elements of A are positive.

It follows from this that the limiting distribution of states is independent of the initial distribution. This surprising result is a direct consequence of the mixing condition, the condition that the a_{ij} be positive. As we shall see, similar results hold for the decision processes discussed below. It is easy to see that this result does not hold in general if we merely assume that $a_{ij} \geq 0$.

11.3 Markovian Decision Processes—I

In place of supposing that the transition mechanism is fixed once and for all by the transition matrix A, let us suppose instead that at each stage we have a choice of the transition matrix to employ. Let us consider two particular processes of this nature, each of some significance in its own right.

In place of A, let us consider a set of matrices $A(q)$ where q runs over some as yet unspecified set of values, scalar or vector. Superimposing this upon the foregoing model of a Markov process, let us consider a multistage decision process in which starting at time $t = 0$ we wish to maximize the probability that S will be in some fixed state, say k, at time $t = T$.

Let

(11.7) $f_i(T) =$ the probability that S is in state k at time T, starting in state i and using an optimal policy.

Then, we readily obtain the relations

(11.8) $f_i(T) = \operatorname*{Max}_{q} \sum_{j=1}^{N} a_{ji}(q) f_j(T - 1), \quad i = 1, 2, \ldots, N, \quad T = 1, 2, \ldots,$

with

(11.9) $$f_i(0) = \delta_{ij}, \quad i = 1, 2, \ldots, N.$$

Note that since we are now counting backwards in time, the coefficients are $a_{ji}(q)$ instead of a_{ij} as before.

11.4 Markovian Decision Processes—II

A process of somewhat different type, which arises in a number of applications, is obtained if we associate with each transition from state j to state i a return, $r_{ij}(q)$, which is dependent upon the decision that is made.

With this addition, we now introduce the problem of determining the sequence of decisions which will maximize the expected return over a T-stage process. Setting

(11.10) $f_i(T) =$ the expected return obtained from a T-stage process, starting in state i and using an optimal policy,

we readily obtain the functional equations

(11.11) $f_i(T) = \operatorname*{Max}_{q} \left[\sum_{j=1}^{N} a_{ji}(q) [r_{ji}(q) + f_j(T - 1)] \right], \quad i = 1, 2, \ldots, N,$

 $f_i(0) = 0.$

As before, a question of interest is that of determining the asymptotic behavior of the function $f_i(T)$ as $T \to \infty$. Of more interest, however, is the question of the nature of the asymptotic optimal policy.

11.5 Steady-state Behavior

It is reasonable to suspect that as $T \to \infty$, the functions $f_i(T)$ will possess asymptotic representations of the following type:

(11.12) $f_i(T) \sim aT + g_i, \quad i = 1, 2, \ldots, N,$

where the scalar a is *independent* of the initial state, and the term g_i is dependent upon i, the initial state.

Under the assumption that all of the transition probabilities are positive, this result can be established quite readily. Furthermore, the method of proof, or the result itself, yields the additional fact that the scalar quantity a and the optimal steady-state policy can be found in the expected fashion, which we shall now describe.

Let us introduce the quantity

$$(11.13) \qquad b_i(q) = \sum_{j=1}^{N} a_{ji}(q) r_{ji}(q),$$

and let $b(q)$ denote the vector whose components are $b_i(q)$, $i = 1, 2, \ldots, N$. Then, letting $A(q) = [a_{ji}(q)]$, and $f(T)$ be the vector whose components are $f_i(T)$, $i = 1, 2, \ldots, N$, we may write (11.11) in the form

$$(11.14) \qquad f(T) = \underset{q}{\text{Max}} \, [b(q) + A(q)f(T-1)].*$$

Here, of course, q denotes a vector, with q_1 the first component, to be chosen in the equation for $f_1(T)$, and so on. Picking a particular policy, q, let us iterate this policy, obtaining for the T-stage return vector the quantity

$$(11.15) \quad f(T; q) = b(q) + A(q)b(q) + A(q)^2 b(q) + \ldots + A(q)^{T-1} b(q).$$

It now follows from standard Markov chain theory that the vector

$$(11.16) \qquad r(q) = \lim_{T \to \infty} f(T; q)/T$$

exists, and, furthermore, that

$$(11.17) \qquad r(q) = a(q) \begin{pmatrix} 1 \\ 1 \\ \cdot \\ \cdot \\ \cdot \\ 1 \end{pmatrix},$$

where $a(q)$ is a scalar quantity, provided that we impose the positivity condition $a_{ji}(q) > 0$.

The policy q yielding the optimal steady-state policy is then the policy which maximizes $a(q)$.

11.6 The Steady-state Equation

Using the asymptotic result, $f(T) = Tr + g$, $r = r(q)$, in 11.14, and dropping the maximization sign, we obtain the system of equations

$$(11.18) \qquad r + g = b(q) + A(q)g.$$

---

* The $A(q)$ is the transpose of the original $A(q)$ mentioned in 11.1. We feel that no harm will be done by dropping the transpose symbol.

This involves the unknown scalar $a = a(q)$ appearing in the components of r, and the N components of g. Apparently, we have N equations for $N + 1$ unknowns. Since, however, $A(q)$ is the transpose of a Markov matrix, its row sum is equal to 1. Hence, if g is a solution, so is g plus any vector of the form of r. Consequently, we may normalize our asymptotic behavior by adding an equation such as

$$(11.19) \qquad \sum_{i=1}^{N} g_i = 1,$$

or, perhaps more simply by taking $g_N = 1$.

11.7 Howard's Iterative Scheme

The previous analytic results can be used to provide an iterative technique for the determination of the optimal steady-state policy. Let $q^{(0)}$ be an initial guess in policy and $g^{(0)}$ be determined by means of the foregoing equations. Let $q^{(1)}$, the next approximation in policy space, be determined by the condition that it yield the maximum over q of

$$(11.20) \qquad b(q) + A(q)g^{(0)} - g^{(0)}.$$

Here, the maximization is performed component by component. Using this new policy, we determine a new r, $r^{(1)}$, and a new g, $g^{(1)}$, and continue in this fashion.

The analytic aspects of this iteration scheme are discussed in Howard's work cited at the end of the chapter, where extensive numerical results may be found, and also in the following chapter where other examples of this approximation in policy space are given.

11.8 Linear Programming and Sequential Decision Processes

Once we have established the nature of the asymptotic behavior of the return functions, and thus the existence of steady-state behavior, it may very well be that various techniques well adapted to static situations, such as linear programming and the theory of games, may yield superior analytic insight and computational approaches.

On the other hand, it can occur that suitable iterative algorithms can be obtained only by regarding the steady-state or static situation as the limit of a dynamic process. The mathematician is perfectly willing to adopt whichever pose yields mastery over the process. This is the great advantage of the abstract formulation, the "as if" attitude that enables the mathematician to focus the powerful techniques of a dozen different mathematical and physical theories upon one particular process.

The steady-state approach can be used with profit in the treatment of the optimal inventory process, at the expense of invoking a theory of positive operators, of which the theory of Markov and positive matrices is only a particular case. Similarly, in the study of "bottleneck processes," or

164

multistage production processes, the asymptotic behavior of positive matrices furnishes the connecting link between the finite stage process treated by dynamic programming techniques and the "expanding economy" model, treated by von Neumann and others.

The steady-state approach can also be employed in the treatment of feedback control processes in which we wish to choose y so as to minimize

$$(11.21) \qquad J(y) = \int_0^T [(x, Bx) + \lambda(y, y)]\, dt,$$

where

$$(11.22) \qquad \frac{dx}{dt} = Ax + y, \quad x(0) = c.$$

On the other hand, one of the most elegant and simple iterative algorithms in the theory of games, the Brown-von Neumann-Robinson technique, uses the fact that the single-stage game can be considered as the steady-state version of a dynamic process.

Similarly, the "flooding" technique of Boldyreff imbeds a static process within a dynamic process. This is characteristic of "relaxation" methods in general.

Bibliography and Comments

11.1 For a detailed discussion of Markov processes, see

W. Feller, *Probability Theory*, John Wiley and Sons, New York, 1948.

For a discussion of Markovian decision processes, see

R. Bellman, *Dynamic Programming*, Princeton Univ. Press, Princeton, N.J., 1957, Chapter XI.

R. Bellman and S. Dreyfus, *Computational Aspects of Dynamic Programming*, Princeton Univ. Press, Princeton, N.J., 1960,

where many computational results will be found, and R. Howard's thesis, cited below.

11.3 This is a stochastic version of the "sequential machine" discussed in Chapter VIII, and perhaps a better model of the medical diagnosis problem. Using the methods of this chapter, it is not difficult to extend the results of Chapter VIII to cover the more general situation of stochastic outcomes.

11.4 Processes of this type arise in various industries. See, for example,

M. Sasieni, *J. British Oper. Res. Soc.*, 1956,

S. Dreyfus, ibid., 1957.

11.5–11.6 Proofs of the statements contained in these sections may be found in

R. Bellman, "A Markovian decision process," *J. Math. and Mech.*, vol. 6, 1957, pp. 679–684.

An introductory account of the theory of positive matrices, together with references to the embracing theory of positive operators, may be found in

R. Bellman, *Introduction to Matrix Analysis*, McGraw-Hill Book Co., Inc., New York, 1960.

11.7 This method is contained in Howard's thesis,

R. Howard, Studies in Discrete Dynamic Programming, PhD Thesis, Mass. Inst. Tech., May, 1958. Since appeared as, R. Howard, "Dynamic Programming and Markov Processes," Technology Press and Wiley, 1960.

There is also a discussion of the much more complicated and realistic situation where some of the coefficients, $a_{ji}(q)$, can be zero.

11.8 For a discussion of a linear programming approach to processes of this nature, see

A. S. Manne, *Linear Programming and Sequential Decision Processes*, Cowles Foundation Discussion Paper No. 62, 1959.

D. Jorgensen and R. Radner, Optimal Scheduling of Checkout and Replacement—III, Checkout and Replacement, 1959, (unpublished).

A number of excellent discussions of input-output analysis and mathematical models of economic systems will be found in

O. Morgenstern, editor, *Economic Activity Analysis*, John Wiley and Sons, New York, 1954.

For the dynamic programming formulation of "bottleneck processes," see Chapters VI and VII of the book *Dynamic Programming*, cited above. The optimal inventory process is treated in Chapter V, and in much greater detail, along with some hydroelectric control processes in

K. J. Arrow, S. Karlin, and H. Scarf, *Studies in the Mathematical Theory of Inventory and Production*, Stanford Univ. Press, Stanford, California, 1958.

For discussion and proof of the Brown-von Neumann-Robinson technique, see

J. Robinson, "An iterative method of solving a game," *Ann. Math.*, vol. 54, 1951, pp. 296–301.

The flooding technique is discussed in

A. Boldyreff, *Determination of the Maximal Steady State Flow of Traffic through a Railroad Network*, The RAND Corporation, Paper P-687, August 5, 1955.

The methods of steepest descent should also be mentioned in this general context. See, for example,

C. B. Tompkins, "Methods of steepest descent," *Modern Mathematics for the Engineer*, edited by E. F. Beckenbach, McGraw-Hill Book Co., Inc., New York, 1956,

as well as the "evolutionary processes" of Box,

G. E. P. Box, "Evolutionary operations; a method for increasing industrial productivity," *Appl. Stat.*, vol. VI, 1957, No. 2, pp. 3–23,

G. E. P. Box and J. S. Hunter, "Condensed Calculations for evolutionary programs," Technical Report No. 27, 1959, Dept. of Math., Princeton University; *Technometrics*, vol. 1, 1959, pp. 77–95.

Finally, let us mention the application of these ideas to prediction theory,

R. E. Kalman and R. W. Koepcke, "Optimal synthesis of linear sampling control systems using generalized performance indexes," *Trans. Amer. Soc. Mech. Eng.*, 1958,

where other references may be found.

CHAPTER XII

QUASILINEARIZATION

Things are seldom what they seem,
Skim milk masquerades as cream.

GILBERT AND SULLIVAN,
"*H.M.S. Pinafore*"

12.1 Introduction

In the preceding chapter devoted to the study of some Markovian decision processes, we encountered some dynamic programming problems which gave rise to the nonlinear vector recurrence relation

$$(12.1) \qquad x(t+1) = \underset{q}{\text{Max}} \, [A(q)x(t) + b(q)].$$

Following the usual steps, which we shall sketch below, we obtain from these equations a nonlinear differential equation of the form

$$(12.2) \qquad \frac{dx}{dt} = \underset{q}{\text{Max}} \, [A(q)x + b(q)].$$

Knowing the nature of the process which generates these equations, we are able to deduce various properties of the solution in simple ways, and, furthermore, we can devise approximation schemes which are far more efficient than those applicable to equations of completely general type.

In view of all this, however, it is natural to ask whether or not these more efficient methods could not be applied to a wider class of differential equations. We shall show that the concept of approximation in policy space, with attendant monotone approximation, can be applied not only to those equations derived directly from decision processes, but also to those equations which can be *interpreted* as describing decision processes.

This is, after all, the essence of the abstract method. Properties of symbols which at first are consequences of the origin of the symbols become eventually properties of the abstract symbols, and thus hold wherever the symbols occur. It is this which makes it so profitable to consider as many different physical settings of the same mathematical equation as possible. Each different stage casts a different light upon the nature of the solution, each illuminates a different phase of its character. Since the total effect is cumulative, the greater the contact with different aspects of reality the greater the understanding of both the abstract and the applied.

12.2 Continuous Markovian Decision Processes

To obtain the differential equation from the difference equation, we proceed as follows. First of all we replace the vector $x(t + 1)$ by the vector $x(t + \Delta)$ and expand

$$(12.3) \qquad x(t + \Delta) = x(t) + \Delta x'(t) + \ldots .$$

Then we replace $r(q)$ by $r(q)\Delta$ and $A(q)$ by $I - A(q)\Delta$. The difference equation now has the form

$$(12.4) \qquad x(t) + \Delta x'(t) = \underset{q}{\text{Max}} \left[\{ I - A(q)\Delta \} x(t) + r(q)\Delta \right] + o(\Delta).$$

Passing to the limit as $\Delta \to 0$ in a purely formal fashion, we obtain the differential equation (12.2).

At this point there are several alternative ways to proceed. We can bluntly *define* a continuous Markov decision process by means of this equation, provided that we show that we have a right to use this nomenclature. This is to say, we must demonstrate that each of the components of $x(t)$ is non-negative for all $t \geq 0$, and that the sum of the components is equal to one for all $t \geq 0$, and, of course, we must establish some existence and uniqueness theorems.

Secondly, we want to show that the limit of the discrete process as $\Delta \to 0$ exists in some sense and is precisely what we have defined as the continuous process.

Alternatively, we can formulate a continuous decision process ab initio and then show that it can be described by means of the differential equation appearing above.

There are advantages and disadvantages to each approach.

12.3 Approximation in Policy Space

Let us now indicate how the technique of approximation in policy space can be employed to treat differential equations like (12.2). Consider initially the scalar equation

$$(12.5) \qquad \frac{du}{dt} = \underset{q}{\text{Max}} \left[b(q, t) + a(q, t)u \right], \quad u(0) = c.$$

Let us guess a policy function $q_0 = q_0(t)$, and determine the corresponding return function $u_0(t)$ by means of the linear equation

$$(12.6) \qquad \frac{du_0}{dt} = b(q_0, t) + a(q_0, t)u_0, \quad u_0(0) = c.$$

168

We next determine $q_1 = q_1(t)$ by the condition that it maximize the function $b(q, t) + a(q, t)u_0$. The associated function $u_1(t)$ is then found by solving the linear equation

$$(12.7) \qquad \frac{du_1}{dt} = b(q_1, t) + a(q_1, t)u_1, \quad u_1(0) = c.$$

Continuing in this way, we determine a sequence of policy functions $\{q_n\}$ and return functions $\{u_n\}$.

It remains to demonstrate convergence, and what is even more important, monotonicity of approximation. Let us remark parenthetically that there are two aspects to iterative techniques. One goal is to obtain an algorithm that leads to an actual solution; the other is to devise methods which will automatically improve any particular approximation.

To demonstrate monotonicity, we proceed as follows. We have the relations

$$(12.8) \qquad \frac{du_1}{dt} = b(q_1, t) + a(q_1, t)u_1, \quad u_1(0) = c,$$

$$\frac{du_0}{dt} = b(q_0, t) + a(q_0, t)u_0, \quad u_0(0) = c$$

$$\leq b(q_1, t) + a(q_1, t)u_0,$$

the last of which holds by virtue of the way in which q_1 was chosen. We wish to conclude that $u_1(t) \leq u_0(t)$ for $t \geq 0$.

To do this, we note that the solution of the linear equation $dv/dt = g(t)v + h(t)$, $v(0) = c$, may be written in the form

$$(12.9) \qquad v = ce^{\int_0^t g(s)\,ds} + \int_0^t e^{\int_s^t g(s_1)\,ds_1} h(s)\,ds,$$

a representation which defines an operator $L(h)$ on the forcing term $h(t)$. Since the exponential is always positive, we see that the inequality $h_1 \geq h_2$ for $t \geq 0$ implies that $L(h_1) \geq L(h_2)$ for $t \geq 0$.

Since the inequality in (12.8) may be written

$$(12.10) \qquad \frac{du_0}{dt} = b(q_1, t) + a(q_1, t)u_0 - p(t),$$

where $p(t) \geq 0$ for $t \geq 0$, it follows that $u_0 \leq u_1$. A straightforward inductive argument now yields the desired inequalities

$$(12.11) \qquad u_n \leq u_{n+1}, \quad n = 0, 1, \ldots, t \geq 0.$$

Since there is no difficulty in showing that the sequence $\{u_n\}$ actually converges to the solution of (12.5), upon imposing the usual conditions on the function $a(q, t)$ and $b(q, t)$, and following the usual procedures, we shall omit the details.

12.4 Systems

Applying precisely the same techniques to the vector equation

$$(12.12) \qquad \frac{dx}{dt} = \text{Max}_q \, [A(q, t)x + b(q, t)], \quad x(0) = c,$$

we see that the proof of monotonicity of approximation depends upon a knowledge of the properties of the functions satisfying the vector inequality

$$(12.13) \qquad \frac{dx}{dt} \geq A(q, t)x.$$

A vector inequality $x \geq y$ is to be understood as equivalent to the N simultaneous inequalities $x_i \geq y_i$, $i = 1, 2, \ldots, N$.

It turns out that it is quite easy to show that the constraint

$$(12.14) \qquad a_{ij}(q, t) \geq 0, \quad i \neq j, t \geq 0,$$

is sufficient to ensure that any function satisfying (12.13) and the relation $x(0) \geq 0$ is actually non-negative for all $t \geq 0$. Fortunately, this is precisely the condition that is automatically satisfied by the equations pertaining to continuous Markovian processes.

12.5 The Riccati Equation

The change of variable $v = u'/u$ converts the general second order linear differential equation

$$(12.15) \qquad u'' + p(t)u' + q(t)u = 0$$

into the nonlinear first order differential equation

$$(12.16) \qquad v' + v^2 + p(t)v + q(t) = 0, \quad v(0) = c.$$

Despite its simple appearance, we know that this equation cannot be solved by means of quadratures when $p(t)$ and $q(t)$ assume general values. Nevertheless, by interpreting (12.16) as an equation of the same form (12.12), we can obtain a representation for the solution of some interest and value.

We begin with the observation that

$$(12.17) \qquad -v^2 = \text{Min}_w \, [w^2 - 2wv].$$

Consequently, (12.16) may be written

$$(12.18) \qquad v' = \text{Min}_w \, [w^2 - 2wv - p(t)v - q(t)], \quad v(0) = c.$$

The discussion given in 12.2 permits us to assert that $v \geq z(w, t)$, where z is the solution of

(12.19) $\qquad z' = w^2 - 2wz - p(t)z - q(t), \quad z(0) = c,$

for some fixed function $w(t)$. Since there is equality if $w = v$, we may write

(12.20) $\qquad\qquad v = \underset{w}{\text{Min }} z(w, t),$

or, using the explicit representation for the solution of (12.19),

(12.21) $\quad v = \underset{w}{\text{Min}} \left[ce^{-\int_0^t [p(s)+2w]\, ds} + \int_0^t [w^2 - q(s)]e^{-\int_s^t [p(s_1)+2w]\, ds_1}\, ds \right].$

This expression may be used to derive various upper bounds for v by taking simple choices for w.

12.6 Extensions

The representation for the function $-v^2$ given in (12.17) is actually only a special case of a result valid for convex functions. If $g''(u) < 0$ for all u, then

(12.22) $\qquad\qquad g(u) = \underset{v}{\text{Max }} [g(v) + (u - v)g'(v)].$

The analytic proof is immediate and the result also follows immediately from a consideration of the geometric interpretation. It follows that a result analogous to (12.21) can be obtained for the equation

(12.23) $\qquad\qquad \dfrac{du}{dt} = g(u) + p(t)u + q(t), \quad u(0) = c.$

12.7 Monotone Approximation in the Calculus of Variations

As we have indicated in Chapter IV, the problem of determining the minimum over y of the functional

(12.24) $\qquad\qquad J(y) = \int_0^T g(x, y)\, dt,$

where

(12.25) $\qquad\qquad \dfrac{dx}{dt} = h(x, y), \quad x(0) = c,$

leads to the nonlinear partial differential equation

(12.26) $\qquad\qquad \dfrac{\partial f}{\partial T} = \underset{v}{\text{Min}} \left[g(c, v) + h(c, v) \dfrac{\partial f}{\partial c} \right].$

As before, we can begin by guessing an initial policy $v_0 = v_0(c, T)$ and then determine f_0 by means of the equation

(12.27)
$$\frac{\partial f_0}{\partial T} = g(c, v_0) + h(c, v_0) \frac{\partial f_0}{\partial c},$$

$$f_0(c, 0) = 0 \quad \text{for all } c.$$

The next policy approximation is obtained by choosing the function $v_1(c, T)$ so as to minimize the function $g(c, v) + h(c, v) \, \partial f_0/\partial c$. The function $f_1(c, T)$ is then obtained as the solution of the equation

(12.28)
$$\frac{\partial f_1}{\partial T} = g(c, v_1) + h(c, v_1) \frac{\partial f_1}{\partial c},$$

$$f_1(c, 0) = 0 \quad \text{for all } c,$$

and so on.

The monotonicity of the sequence $\{f_n\}$ depends upon the properties of the solutions of the partial differential inequality

(12.29)
$$\frac{\partial f}{\partial T} - h(c, v_1) \frac{\partial f}{\partial c} \geq 0, \quad f(c, 0) \geq 0.$$

It is easy to show that this implies that $f \geq 0$ for $T \geq 0$, and thus that $f_0 \geq f_1 \geq \cdots \geq f_n \geq f_{n+1} \geq \cdots$.

The problem of convergence of the sequence is, however, now much more complex.

12.8 Computational Aspects

A reasonable way to proceed to determine a solution of the equation $g(t) = 0$, given the information that t_0 is an approximate solution, is to use a straight line approximation to the function $g(t)$ in the neighborhood of $t = t_0$. Thus, we write

(12.30)
$$0 = g(t) = g(t_0) + (t - t_0)g'(t_0) + \cdots,$$

thereby obtaining as a further approximation to the solution, the value

(12.31)
$$t_1 = t_0 - \frac{g(t_0)}{g'(t_0)}.$$

Continuing in this way, we obtain the iterative scheme

(12.32)
$$t_{n+1} = t_n - \frac{g(t_n)}{g'(t_n)}.$$

This is Newton's method of approximation, a technique which possesses not only the merit of simplicity, but also the property that the error at any stage is approximately the square of the error at the preceding stage.

Thus, instead of convergence of the order of r^n at the n-th stage, for some $r < 1$, we have convergence of the order of r^{2^n}.

Combining this idea with those described above, we are led to use the algorithm

$$(12.33) \qquad \frac{du_n}{dt} = g(u_{n-1}, t) + (u_n - u_{n-1})g_u(u_{n-1}, t), \quad u_n(0) = c,$$

to generate the solution of

$$(12.34) \qquad \frac{du}{dt} = g(u, t), \quad u(0) = c,$$

starting with an initial guess $u_0(t)$. Not only does this technique yield very rapid solution, but in a number of cases, we have the bonus of monotone convergence.

12.9 Two-point Boundary-value Problems

Quasilinearization techniques yield a new approach to the study of two-point boundary-value problems. They enable us to obtain the solution of a nonlinear equation using solutions to a succession of linear equations.

Let us begin by reviewing the method used to solve two-point boundary-value problems for linear differential equations. It is sufficient for our purposes to consider the second-order scalar equation

$$(12.35) \qquad u'' + p(t)u' + q(t)u = f(t),$$
$$u(0) = c_1, \quad u(T) = c_2,$$

where $f(t)$ may or may not be identically zero. In any case, we suppose that it is known that this problem has a unique solution, and our task is merely to determine it.

Let $u_1(t)$ be the solution of the homogeneous equation

$$(12.36) \qquad u_1'' + p(t)u_1' + q(t)u_1 = 0,$$
$$u_1(0) = 0, \quad u_1'(0) = 1,$$

an initial value problem, and $u_2(t)$ be the solution of

$$(12.37) \qquad u_2'' + p(t)u_2' + q(t)u = f(t),$$
$$u_2(0) = 1, \quad u_2'(0) = 0,$$

another initial value problem.

To obtain the solution of (12.35), we set

$$(12.38) \qquad u = c_1 u_2 + a_1 u_1,$$

where the parameter a_1 is to be determined by the condition at $t = T$.

173

Observe that the condition at $t = 0$ is automatically satisfied. We have

(12.39)
$$c_1 u_2(T) + a_1 u_1(T) = c_2.$$

That $u_1(T) \neq 0$ is a consequence of our assumption of the uniqueness of the solution of (12.35). Hence,

(12.40)
$$a_1 = [c_2 - c_1 u_2(T)]/u_1(T).$$

Let us now turn to the equation

(12.41)
$$u'' + g(u, u') = 0,$$
$$u(0) = c_1, \ u(T) = c_2.$$

Let $u_0(t)$ be an initial guess and let the subsequent approximations be determined by means of the relation

(12.42) $\quad u_n'' + g(u_{n-1}, u_{n-1}') + p_n(t)(u_n - u_{n-1}) + q_n(t)(u_n' - u_{n-1}') = 0,$

where

(12.43)
$$p_n(t) = \frac{\partial g}{\partial u}(u, u'), \quad q_n(t) = \frac{\partial g}{\partial u'}(u, u'),$$

both evaluated at $u = u_{n-1}$, $u' = u_{n-1}'$.

The questions of convergence of the sequence $\{u_n(t)\}$ and monotonicity can be answered in terms of the positivity properties of the linear equation

(12.44)
$$u'' + q_n(t)u' + p_n(t)u = 0.$$

These matters are discussed in the work of Kalaba referred to at the end of the chapter.

12.10 Partial Differential Equations

In order to apply similar techniques to the study of the nonlinear elliptic equation, such as

(12.45)
$$u_{xx} + u_{yy} = e^u,$$

or to a nonlinear parabolic equation such as

(12.46)
$$u_t = u_{xx} + u^2,$$

it is necessary to study the linear equations

(12.47)
$$u_{xx} + u_{yy} + q(x, y)u = v(x, y),$$
$$u_t - u_{xx} + q(x, y)u = v(x, y).$$

The positivity properties of the solutions of these equations, with suitable boundary conditions, are dependent upon the properties of the associated Green's functions.

174

Fortunately, the fundamental equations of mathematical physics possess the required positivity properties, enabling us to present algorithms which yield monotone convergence to the solution of equations such as (12.45) or (12.46).

References to some of the work done in this connection will be found below.

Bibliography and Discussion

12.1 A detailed discussion of quasilinearization, with both theoretical and computational results, may be found in

R. Kalaba, "On nonlinear differential equations, the maximum operation, and monotone convergence," *J. Math. and Mech.*, vol. 8, 1959, pp. 519–574.

Earlier results may be found in

R. Bellman, "Functional equations in the theory of dynamic programming —V: positivity and quasilinearity," *Proc. Nat. Acad. Sci. USA*, vol. 41, 1955, pp. 743–746.

————, *Dynamic Programming*, Princeton Univ. Press, Princeton, New Jersey, 1957.

12.2–12.6 These processes were first discussed in the book cited above in which the results contained here were presented. The second path referred to in this section is followed by T. E. Harris in his monograph,

T. E. Harris, *Branching Processes*, Ergebnisse der Math., 1960,

where many continuous stochastic processes are studied. This approach requires deep, subtle and determined analysis.

12.7 This method was first given in

R. Bellman, "Monotone approximation in dynamic programming and the calculus of variations," *Proc. Nat. Acad. Sci. USA*, vol. 40, 1954, pp. 1073–1075.

12.8–12.9 A discussion of Newtonian convergence, together with favorable examples, will be found in the paper by Kalaba cited above.

12.10 Detailed discussion of positive operators will be found in the above cited paper by Kalaba, in the monograph

E. F. Beckenbach and R. Bellman, *Inequalities*, Ergebnisse der Math., 1960,

and in the monograph

C. A. Caplygin, *New Methods in the Approximate Integration of Differential Equations*, (Russian), Leningrad, 1950.

where upper and lower bounds are deduced for the solutions of linear and nonlinear differential equations.

CHAPTER XIII

STOCHASTIC LEARNING MODELS

Ask, and it shall be given you;
Seek, and ye shall find;
Knock, and it shall be opened unto you.

Matthew VII., 7.
New Testament

13.1 Introduction

Partway along the road connecting the mathematical models of stochastic decision processes already discussed and the adaptive control processes we shall discuss subsequently are the stochastic learning models of Bush and Mosteller and those of Flood. Since these are descriptive processes, in the sense that a certain mechanism is postulated, rather than decision processes in which this mechanism is derived on the basis of optimization, we shall dwell only very briefly at this way-station. The reader interested in these matters may refer to the book by Bush and Mosteller, cited at the end of the chapter, and a number of papers by Flood, which may also be found there.

13.2 A Stochastic Learning Model

In order to construct a mathematical model of how an organism gradually adapts its behavior to the information that it receives in an environment possessing certain strange and unknown features, we proceed as follows. Consider a multistage process in which N alternative actions, $A_1, A_2, \ldots,$ A_N are possible at each stage, and suppose that initially the organism possesses a certain probability distribution $p = (p_1, p_2, \ldots, p_N)$ which determines the probability with which it will choose any particular alternative at the initial stage.

Let us assume to begin with that a choice of an alternative, say A_i, always leads to a particular consequence dependent upon the alternative that is selected. We further suppose that this consequence results in a change in the preferences that the organism has for different alternatives. To indicate this change in preferences, we transform the original probability distribution, p, into a new distribution, p', where

$$(13.1) \qquad p' = T_i(p) = [T_{i1}(p), T_{i2}(p), \ldots, T_{iN}(p)].$$

Since p' must represent a probability distribution, we must have the following relations satisfied by the components of $T_i(p)$,

$$(13.2) \quad (a) \quad T_{ij}(p) \geq 0,$$

$$(b) \quad \sum_{j=1}^{N} T_{ij}(p) = 1,$$

for each value of i.

The process now continues in this fashion. Observe that as formulated it is a Markovian model, in that only the current state of the system, the probability distribution, is of any interest, not the past history of the system.

13.3 Functional Equations

As is usually the case in stochastic processes, what is of interest is the ultimate behavior of the system. Let us introduce the following functions:

$$(13.3) \qquad f_i(n, p) = \text{the probability that at stage } n \text{ the } i\text{-th}$$
alternative will be chosen, given the
initial probability distribution of
alternatives, p, $i = 1, 2, \ldots, N$.

Then the usual argument yields for each i the functional equation

$$(13.4) \qquad\qquad f_i(n, p) = \sum_{j=1}^{N} p_j f_i[n - 1, T_j(p)],$$

with the initial condition

$$(13.5) \qquad\qquad f_i(0, p) = p_i, \quad i = 1, 2, \ldots, N.$$

13.4 Analytic and Computational Aspects

There are a number of quite interesting analytic problems connected with this apparently simple functional equation, which we shall not go into here. Although this recurrence relation can be used to compute the sequence $\{f_i(n, p)\}$ in a straightforward way, the convergence is rather slow. A number of elegant analytic representations can be obtained for the solutions, in different regions of p-space, which yield much more rapid numerical estimation.

13.5 A Stochastic Learning Model—II

Closer to the sequential machine discussed in Chapter VIII, and to the stochastic decision processes discussed in Chapter X, is a process in which a choice of A_i results in a distribution of outcomes, rather than any fixed outcome. Let us suppose then that a choice of A_i results in a possible set of outcomes $O_{i1}, O_{i2}, \ldots, O_{iM}$, with an associated conditional probability

177

distribution $[q_{i1}, q_{i2}, \ldots, q_{iM}]$. If A_i is chosen and O_{ij} occurs, then we transform p into a new probability distribution $T_{ij}(p)$. Retaining the same definition for the function $f_i(n, p)$, we see that the analogue of the recurrence relation of (13.4) is

$$(13.6) \qquad f_i(n, p) = \sum_{j=1}^{N} p_j \left\{ \sum_{k=1}^{M} q_{jk} f_i[n - 1, T_{jk}(p)] \right\}.$$

13.6 Inverse Problem

As mentioned above, there are a number of interesting analytic problems associated with functional equations of this type—equations which are natural extensions of those derivable from the conventional Markov chains.

Since, however, we actually wish to compare the results of experiments on rats, dogs, and humans with the results predicted by the equations, we face the additional problems involved in determining the nature of the transformations $T_{ij}(p)$ from experimental data, and of deciding whether or not the processes described above actually correspond to the learning processes observed in the laboratory.

These are very complex questions, compounded of a mixture of analysis, statistics, and faith. For reference to the corresponding inverse problem for deterministic processes, see 1.3 and the discussion at the end of that chapter.

Bibliography and Discussion

13.1 See the book

R. R. Bush and F. Mosteller, *Stochastic Models for Learning*, John Wiley and Sons, New York, 1955,

and a later paper

M. I. Hanania, "A generalization of the Bush-Mosteller model with some significance tests," *Psychometrika*, March, 1959, pp. 53–66.

Let us also cite the following papers of M. Flood, which furnish a cross-section of the work he has done in this field.

M. Flood, *A Preference Experiment*, The RAND Corporation, Paper P-256, November 13, 1951.

————, *An Experimental Multiple Choice Situation*, The RAND Corporation, Paper P-313, 1952.

————, "A stochastic model for social interaction," *Trans. N.Y. Acad. Sci.*, ser. II, vol. 16, 1954, pp. 202–205.

————, "Game-learning theory and some decision-making experiments," *Decision Processes*, John Wiley and Sons, New York, 1954, pp. 139–158.

Other references will be found in these sources. See also the interesting expository work,

W. Ross Ashby, *An Introduction to Cybernetics*, John Wiley and Sons, New York, 1954.

13.4 A detailed discussion of analytic and computational aspects of an interesting version of this general equation may be found in

R. Bellman, T. E. Harris, and H. N. Shapiro, *Studies in Functional Equations Occurring in Decision Processes*, The RAND Corporation, Research Memorandum RM-878, 1952,

and further results are given in

S. Karlin, "Some random walks arising in learning processes, I," *Pacific J. Math.*, vol. 3, 1953, pp. 725–756.

An interesting paper in this field is

S. Gorn, "On the mechanical simulation of habit-forming and learning," *Information and Control*, vol. 2, 1959, pp. 226–259.

CHAPTER XIV

THE THEORY OF GAMES AND PURSUIT PROCESSES

A man said to the universe
"Sir, I exist!"
"However," replies the universe
"The fact has not created in me
A sense of obligation."

STEPHEN CRANE
"*A Man Said to the Universe*"

14.1 Introduction

In the preceding chapters where stochastic control processes were discussed, we encountered physical systems ruled by recurrence relations of the form

(14.1) $$x_{n+1} = g(x_n, y_n, r_n), \quad x_0 = c.$$

Here, x_n represents the state of the system at time n, r_n is a stochastic vector, and y_n represents the control vector.

On the assumption that the distribution function for the independent random variables, $\{r_n\}$, is known, and on the further premise that we are willing to compromise our objectives by minimizing or maximizing the expected value of a criterion function, we were able to formulate a theory of control processes in the presence of stochastic effects. Furthermore, it was shown that this could be done using precisely the same formalism employed in the study of deterministic control processes.

A valid objection that must be made against the theory presented in Chapter X is that it is not always the case that the probability distribution for r_n is known. The present chapter in which we take this into account must then be viewed as a halfway house between the domain of stochastic control processes involving *known* distribution functions and those involving *unknown* distribution functions. These latter, more recondite processes, will be discussed in Chapters XV–XVII.

At the moment, we wish to meet the objection made above by means of a quite interesting and novel approach based upon the concept of a "game against nature." To fix the unknown probability distribution, we shall suppose that it is chosen by outside forces (sometimes called Gremlins, sometimes less printable epithets) in the most disadvantageous way possible.

Oddly enough, it is far easier to construct a mathematical theory for a situation of this type, complete opposition, than in the case where this is only partially the case.

The mathematical setting for the approach we shall pursue below is the *theory of games*, created independently by Borel and von Neumann. In order to set the stage for the reader who may not be acquainted with this theory, we shall present the basic ideas of the theory of finite single-stage games. Following this, we shall indicate how the theory of multistage games may be approached by means of the functional equation technique of dynamic programming. Finally, we shall illustrate the applicability of these techniques to the study of stochastic control processes.

A particularly interesting class of processes of this nature go under the heading of "games of survival," of which an important subclass are the pursuit processes.

14.2 A Two-person Process

Beginning on what appears to be a rather simple level, consider the following situation involving two individuals, or "players," unimaginatively called A and B. The first player, A, has a choice of M different decisions, or "moves," which we designate by means of the numbers $1, 2, \ldots, M$, and the second player has a choice of N different alternatives, $j = 1, 2, \ldots, N$.

If A makes the i-th choice and B makes the j-th choice, A receives a quantity a_{ij} and B a quantity b_{ij}. The matrices (a_{ij}) and (b_{ij}) are called *payoff matrices*.

In a situation of this type, the problem of determining "optimal play" clearly depends upon the conventions of play and the criteria of each player. Let us consider the simple case in which A wishes to maximize his return and B wishes, likewise, to maximize the return that he receives.

If A is forced to make his choice before B chooses j, then B's choice of j will be dictated by the requirement that it maximize the quantity b_{ij}. This determines j as a function i, $j(i)$, the maximizing index for B.

With j fixed in this way, it is clear that A will choose i so as to maximize the quantity

(14.2) $$a_{i,j(i)}.$$

A similar situation results if B is forced to move before A.

14.3 Multistage Process

Let us now consider, within this framework, a multistage control process in which an opponent attempts to choose the stochastic vector so as to maximize a minimum deviation, or minimize a maximum gain.

Referring to our customary recurrence relation,

(14.3) $$x_{n+1} = g(x_n, y_n, r_n), \quad x_0 = c,$$

let us consider a multistage process in which A wishes to minimize the expected value of $h(x_N)$ and B wishes to maximize this expression. We envisage a feedback control process which proceeds along the following lines. A chooses a control vector y_1, subject to restrictions on the magnitude of the components, and B, with full knowledge of this choice of A's, chooses a vector r_1, subject also to restrictions upon the components.

For a one-stage process, y_1 and r_1 will be chosen in accordance with the expression

(14.4) $$\text{Min Max } h[g(c, y_1, r_1)].$$
$$ y_1 r_1$$

Observe that while A chooses a *quantity*, y_1, B must choose a *function*, $r_1 = r_1(y_1)$, namely the value of r_1 which yields the maximum in (14.4). Calling this min-max value $f_1(c)$, we see that the multistage continuation of this process, with alternate choices by A and B, leads to the recurrence relation

(14.5) $$f_N(c) = \text{Min Max} [f_{N-1}[g(c, y_1, r_1)]].$$
$$ y_1 r_1$$

Once again, the feedback aspect considerably simplifies the mathematical problem. Were A and B to choose the values y_i and r_i, $i = 1, 2, \ldots, N$, all at the same time, we would face the hopelessly difficult problem of determining the value of

(14.6) $$\text{Min} \qquad \text{Max} \qquad h(x_N).$$
$$[y_1, y_2, \ldots, y_N] \; [r_1, r_2, \ldots, r_N]$$

We leave it to the reader to convince himself that this problem cannot, in general, be reduced to the recurrence relation of (14.5).

14.4 Discussion

In a similar fashion, we obtain corresponding recurrence relations governing the process in which B is forced to make his decision before A's. These results can then be used to furnish upper and lower bounds for the outcome of processes in which A and B are forced to make their moves simultaneously, independent of the knowledge of the other's choice.

It turns out that the situation becomes very much more interesting from both the physical and mathematical point of view if we force the players to move simultaneously.

14.5 Borel-von Neumann Theory of Games

Let us now sketch the fundamental idea of the theory of games, invented independently by Borel and von Neumann. In order to overcome the difficulty involved in this concept of simultaneous moves, we replace the idea of a definite choice by that of a probability distribution of choices.

Once again, we see that a suitable generalization of a state is that of a probability distribution of states.

In place of requiring that A make a definite choice, i, we ask that he furnish a probability distribution $[p_1, p_2, \ldots, p_N]$ for these choices, and that B likewise choose a probability distribution $[q_1, q_2, \ldots, q_M]$. Here p_i is the probability that A makes the i-th move, and q_j the probability that B makes the j-th move.

Having agreed to accept probability distributions in lieu of precise choices, we are in some measure bound to evaluate outcomes in terms of expected values. The expected return to A will be the expression

$$(14.7) \qquad E_A(p, q) = \sum_{i,j} a_{ij} p_i q_j$$

with

$$(14.8) \qquad E_B(p, q) = \sum_{i,j} b_{ij} p_i q_j$$

the expected return for B.

It would appear that we have gained very little from this complication of our process by the introduction of stochastic aspects, but, as we shall see, this is not the case. It is, however, true that we must relinquish some of the foregoing generality if we wish to obtain fruitful results.

It turns out that a wholly satisfactory theory exists only in the case where

$$(14.9) \qquad b_{ij} = -a_{ij}.$$

Henceforth, we shall make this assumption.

Two-person processes of the type we have just described are called *"games,"* which is a bit misleading since some of the most serious aspects of our civilization can be described in these terms. If the foregoing assumption holds, the game is called *zero sum*; if not, it is called *non-zero sum*.

Processes of this general nature can also be formulated for the case where there are more than two individuals involved. These are called *n-person games*. Despite the obvious interest in processes of this kind, their study languishes under a cloud of indefiniteness. No definitive theory exists for these n-person games, corresponding to the results presented below for the two-person zero-sum case.

Let us assume that A wishes to maximize his expected return and that B wishes to maximize his expected return. In view of our zero-sum hypothesis, this is equivalent to B wishing to minimize A's expected return.

If A is forced to choose his probability distribution before B, the expected return for A will have the form

$$(14.10) \qquad \operatorname*{Max}_{[p]} \operatorname*{Min}_{[q]} E_A(p, q).$$

183

If B is forced to make his choice of a probability distribution before A, the expected return to A will be

$$(14.11) \qquad \underset{[q]}{\text{Min}} \ \underset{[p]}{\text{Max}} \ E_A(p, q).$$

In each case, the variation is taken over the regions determined by the relations

$$(14.12) \quad (a) \quad p_i \geq 0, \quad \sum_{i=1}^{N} p_i = 1,$$

$$(b) \quad q_j \geq 0, \quad \sum_{j=1}^{M} q_j = 1.$$

We do not seem to have advanced in any direction, save that of complexity, since we still seem to have to consider the case of one decision ahead of the other. But wait!

14.6 The Min-max Theorem of von Neumann

The keystone of the theory of games is the following remarkable result of von Neumann, established previously by Borel only for the case where $M, N \leq 4$:

$$(14.13) \qquad \underset{[p]}{\text{Min}} \ \underset{[q]}{\text{Max}} \ E_A(p, q) = \underset{[q]}{\text{Max}} \ \underset{[p]}{\text{Min}} \ E_A(p, q).$$

The common value of these two expressions is called the *value* of the game. Occasionally, to show its dependence upon the payoff matrix, we write $v = v(a_{ij})$.

14.7 Discussion

This result has the following implication. A can announce his probability distribution in advance without giving B the slightest advantage, and B can do likewise. This is certainly an astonishing fact. By means of this mathematical artifice, the introduction of a choice of a probability distribution, we have eliminated the difficulties of simultaneity.

We shall not enter here into any defense of the philosophical premises of the theory of games. Whatever the conceptual difficulties, as a mathematical theory it is of great interest and significance and contains many fascinating regions as yet little explored. The reader who is interested in examining the foundations may refer to a number of references to be found at the end of the chapter.

As in the case of stochastic control processes, a number of difficulties arise when we attempt to fit a physical process into this mathematical mold. Once, however, we have overcome our qualms and have agreed to do so, it is then an easy matter, as we shall see below, to use the functional equation technique of dynamic programming to study various types of multistage decision processes of this character.

184

14.8 Computational Aspects

It is not a simple matter to compute the value of a game if the matrix (a_{ij}) is of high dimension. There are, however, a number of algorithms which are suitable for machine computation. One of these is based upon linear programming techniques.

Generally speaking, the most efficient algorithms are those which take account of the fact that what is called the value of a game is actually a static quantity associated with the steady-state equilibrium process of a dynamic process. Algorithms due in principle to Brown and von Neumann, and brought to full flower by Julia Robinson, H. N. Shapiro, and J. Kemeny, exploit this fact.

14.9 Card Games

It is natural to expect that a theory of this type would be applied to some of the more interesting social diversions, such as bridge and poker. A number of mathematical models of poker have been investigated, with results that coincide with our intuitive feelings concerning various betting policies, bluffing, and so on.

The actual game of poker, involving many players, with individual utilities, and many delicate psychological overtones, is far outside the reach of any mathematical analysis. Conceivably, the day may come when games of this type can be analyzed precisely. If so, this success will be due not so much to any perfection of mathematical techniques, but to the sad fact that at that far off time, conformity will have taken its inexorable toll to the extent that human beings will not exhibit the individuality that they do in these days.

14.10 Games of Survival

What we have described above in 14.5 and 14.6, is a single-stage game. Preliminary to a discussion of control processes and pursuit processes, let us consider *"games of survival."*

In addition to the factors we have already included, let us add that of finite fortunes. Suppose that A possesses a quantity u initially and that B possesses v. The game process described above in 14.2 continues until either A or B is ruined. This is a version of the classical "gambler's ruin" problem, in which the players determine their transition probabilities at each stage.

In order to treat a multistage process of this type, we call once again upon the functional equation technique. Let us introduce the function

(14.14) $f(u, v) =$ the probability that A will survive B when
A has $u > 0$, B has $v > 0$, and both
employ optimal policies.

185

Using the principle of optimality, we readily obtain the functional equation

(14.15) $\quad f(u, v) = \underset{p}{\text{Max}} \underset{q}{\text{Min}} \left[\sum_{i,j} p_i q_j f(u + a_{ij}, v - a_{ij}) \right],$

$\qquad\qquad = \underset{q}{\text{Min}} \underset{p}{\text{Max}} \left[\qquad \cdots \qquad \right],$

$\qquad\qquad u > 0, \quad v > 0,$

with the boundary conditions

(14.16) $\qquad\qquad f(u, v) = 0, \quad u \leq 0, \quad v > 0,$

$\qquad\qquad\quad f(u, v) = 1, \quad u > 0, \quad v \leq 0.$

Since the game is zero-sum, we can simplify matters to some extent by writing

(14.17) $\qquad\qquad\qquad f(u) \equiv f(u, v).$

Then the equation in (14.15) can more simply be written

(14.18) $\quad f(u) = \underset{p}{\text{Max}} \underset{q}{\text{Min}} \sum p_i q_j f(u + a_{ij}), \quad 0 < u < c,$

where c is the total amount of money in the game. The boundary conditions are now

(14.19) $\qquad\qquad\qquad f(u) = 0, \quad u \leq 0,$

$\qquad\qquad\qquad\quad f(u) = 1, \quad u \geq c.$

The study of equations of this type is rather more complex than those previously encountered, since the question of uniqueness of solution is not at all trivial. Observe that these are matters of some interest if we wish to use these equations for computational purposes.

14.11 Control Processes as Games Against Nature

We are now ready to discuss control processes subject to random effects in the context of "*games against nature.*" In the recurrence relation

(14.20) $\qquad\qquad\qquad x_{n+1} = g(x_n, y_n, r_n),$

let us agree to regard y_n as a vector at our disposal and r_n as a vector at the disposal of an adversary. This fictitious adversary we can regard as *Nature*. The fact that we regard nature as an opponent is a manifestation of the usual anthropomorphic tendencies.

Consider, in this light, the problem of terminal control. We wish to choose the y_n so as to minimize the expected value of a prescribed function of the terminal position, while nature wishes to maximize this expected value.

Let us suppose that we are allowed M different choices for y_1, say $y_1^{(1)}, \ldots, y_1^{(M)}$, and similarly that Nature is permitted N different choices for r_1, say $r_1^{(1)}, \ldots, r_1^{(N)}$.

This is consistent with our usual desire to avoid any discussion of continuous stochastic processes. As before, the moves are made in the following order: first y_1 and r_1 at the first stage, then y_2 and r_2 at the second stage, and so on. By this, we mean first a choice of the probability distributions for y_1 and r_1, then a choice of the probability distributions for r_2 and y_2, and so on.

Considering this as a two-person multistage game, let $p = [p_1, p_2, \ldots, p_M]$ denote the probability distribution for y_1 and $q = [q_1, q_2, \ldots, q_N]$ the probability distribution for r_1.

Let $h(x_k)$ be the criterion function measuring the utility of the state at the end of a k-stage process. Introduce the function

$$(14.21) \qquad f_k(c) = \underset{\{p^i\}}{\text{Min}} \underset{\{q^j\}}{\text{Max}} \; \text{Exp} \; h(x_k),$$

where the minimum is taken over the distributions chosen at each stage by us, the maximum is taken over the distributions chosen at each stage by Nature, and the choices are made in the manner indicated above. Then, applying the principle of optimality in conjunction with the min-max theorem of 14.6, we see that

$$(14.22) \qquad f_k(c) = \underset{p}{\text{Min}} \underset{q}{\text{Max}} \left[\sum_{i,j} p_i q_j f_{k-1}[g(c, y_1^{(i)}, r_1^{(j)})] \right],$$

$$= \underset{q}{\text{Max}} \underset{p}{\text{Min}} \left[\qquad \ldots \qquad \right],$$

$k = 2, 3, \ldots,$

$$(14.23) \qquad f_1(c) = \underset{p}{\text{Min}} \underset{q}{\text{Max}} \left[\sum_{i,j} p_i q_j h[g(c, y_1^{(i)}, r_1^{(j)})] \right],$$

$$= \underset{q}{\text{Max}} \underset{p}{\text{Min}} \left[\qquad \ldots \qquad \right].$$

Once again, we have a computational algorithm quite analogous to that obtained for both deterministic and stochastic control processes. The time required to obtain a numerical solution is now strongly dependent upon the dimensionality of p and q, since, as mentioned above, the determination of both the value and optimal strategies for an $M \times N$ game is not simple if M and N are large. If, however, M and N are both less than or equal to 5 or 10, we may consider the computational solution well within our grasp, provided, of course, that the dimension of c is not too large.

14.12 Pursuit Processes—Minimum Time to Capture

"Suppose that we have two individuals, P, the pursuer, and Q, the quarry, moving in the plane, or in space. The pursuer wishes to capture the quarry, and the quarry wishes to avoid capture. What are the paths for each?"

187

These problems are considerably more vexing to solve, and even to formulate, than one might suspect, since they fall within the domain of multistage games of continuous type. Here we shall treat them by means of the theory of multistage games, allowing only discrete processes where P and Q can go only from one lattice point of a preassigned grid to one of a neighboring set of lattice points, at discrete times.

Let us then consider the grid of points $[m\delta, n\delta]$, $-N \le m, n \le N$, with P and Q both constrained to be at one of these lattice-points.

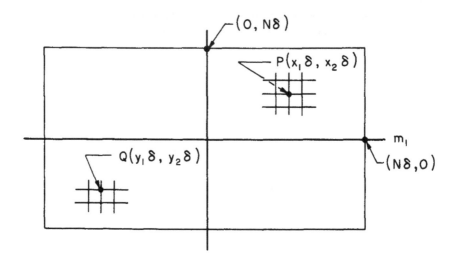

Figure 20

Furthermore, we make the process discrete in time, moves taking place at times $0, 1, \ldots$, and simultaneously. At any particular time, P can move to any lattice point within a square of length $2v_1\delta$ surrounding his current position as center, unless the boundary of the region intervenes. In this case, he can go only as far as a lattice point on the boundary. Similarly, Q can move to any lattice point within a square of length $2v_2\delta$ surrounding his current position as center, subject to the same constraint.

Let us now suppose that P moves so as to minimize the expected time to capture and that Q moves so as to maximize this expected time. We then introduce the function

(14.24) $f(x_1, x_2; y_1, y_2) =$ the expected time to capture when P is at $(x_1\delta, x_2\delta)$, Q is at $(y_1\delta, y_2\delta)$, and both P and Q employ optimal policies.

188

On the first move, P will go to one of the allowable points $(x_1 + i\delta, x_2 + j\delta)$ with probability p_{ij}, while Q will jump to one of his allowable points, $(y_1 + k\delta, y_2 + \ell\delta)$, with probability $q_{k\ell}$. We thus obtain the functional equation

(14.25)
$$f(x_1, x_2; y_1, y_2) = \underset{[p_{ij}]\,[q_{k\ell}]}{\text{Min Max}}\left[\sum p_{ij}q_{k\ell}f(x_1 + i\delta, x_2 + j\delta; y_1 + k\delta, y_2 + \ell\delta)\right]$$
$$= \underset{[q_{k\ell}]\,[p_{ij}]}{\text{Max Min}}\left[\qquad\qquad\ldots\qquad\qquad\right].$$

It is expected that as $\delta \to 0$ the solution of this functional equation will approach that of the continuous pursuit process. Actually, a more logical approach is to define the solution of the continuous process in terms of the limit of the discrete process described above. Although a little work has been done in this direction, much remains to be done.

14.13 Pursuit Processes—Minimum Miss Distance

If we have no a priori certainty that capture will ensue, we can study instead the minimum miss distance. If this turns out to be zero, then capture actually occurs.

Let

(14.26) $\quad f(x_1, x_2; y_1, y_2) =$ the expected minimum distance ever
separating P and Q when P starts at
$(x_1\delta, x_2\delta)$, Q at $(y_1\delta, y_2\delta)$, and both
employ optimal policies.

It is easily seen that

(14.27)
$$f(x_1, x_2; y_1, y_2) = \underset{[p_{ij}]\,[q_{k\ell}]}{\text{Min Max}}\left[\sum p_{ij}q_{k\ell} \text{ Min }\{[(x_1 - y_1)^2 + (x_2 - y_2)^2]^{1/2},\right.$$
$$\left. f(x_1 + i\delta, x_2 + j\delta; y_1 + k\delta, y_2 + \ell\delta)\}\right],$$
$$= \underset{[q_{k\ell}]\,[q_{ij}]}{\text{Max Min}}\left[\qquad\qquad\ldots\qquad\qquad\right],$$
$$= \text{Min }\left[[(x_1 - y_1)^2 + (x_2 - y_2)^2]^{1/2},\right.$$
$$\left.\underset{[p_{ij}]\,[q_{k\ell}]}{\text{Min Max}}\left[\sum p_{ij}q_{k\ell}f(x_1 + i\delta, x_2 + j\delta; y_1 + k\delta, y_2 + \ell\delta)\right]\right].$$

14.14 Pursuit Processes—Minimum Miss Distance within a Given Time

In order to solve these equations computationally, we can either use approximation in function space, the usual method of successive approximations, or approximation in policy space. One way of approximating, a technique we have employed before, is to introduce a finite time, T, for the duration of the process. Letting

(14.28)

$f_T(x_1, x_2; y_1, y_2) =$ the expected minimum distance separating
P and Q when P starts at $(x_1\delta, x_2\delta)$, Q at
$(y_1\delta, y_2\delta)$ and both employ optimal
policies over the time interval $[0, T]$.

The basic functional equation is

(14.29)
$$f_T(x_1, x_2; y_1, y_2) = \underset{[p_{ij}]}{\text{Min}} \underset{[q_{k\ell}]}{\text{Max}} \left[\sum p_{ij}q_{k\ell} \text{Min} \left\{ [(x_1 - y_1)^2 + (x_2 - y_2)^2]^{1/2}, \right. \right.$$
$$\left. \left. f_{T-\Delta}(x_1 + i\delta, x_2 + j\delta; y_1 + k\delta, y_2 + \ell\delta) \right\} \right]$$
$$= \underset{[q_{k\ell}]}{\text{Max}} \underset{[p_{ij}]}{\text{Min}} \left[\quad \ldots \quad \right].$$

14.15 Discussion

In what has preceded, we have sketched an application of the theory of games to the study of various stochastic control processes. This approach has been of interest in enabling us to circumvent the rather awkward assumption that the statistics of the random effects are known.

On the other hand, we certainly cannot take too seriously a theory that assumes that Nature is always directly opposed to our interests. We say this despite the well-known concept of the *"perversity of inanimate objects."**

This theory is, however, extremely useful in a number of cases since it provides lower bounds or upper bounds to performance. Furthermore, it is very stimulating conceptually, and leads to many elegant analytic results.

In what follows, we shall use an entirely different approach to the problem of unknown effects, one that is more suited to the multistage character of feedback control.

Bibliography and Discussion

14.1–14.7 An excellent introduction to the theory of games is

J. D. Williams, *The Compleat Strategyst*, McGraw-Hill Book Co., Inc., New York, 1955.

The classic work is

J. von Neumann and O. Morgenstern, *The Theory of Games and Economic Behavior*, Princeton Univ. Press, Princeton, N.J., 1948.

---

* This concept is by now fairly well established by means of a number of careful experiments. Perhaps the most conclusive involved dropping a piece of buttered toast on a rug. In 79.3% of the trials, the toast fell buttered side down.

BIBLIOGRAPHY AND DISCUSSION

The most recent authoritative account is

R. Duncan Luce and H. Raiffa, *Games and Decisions, Introduction and Critical Survey*, John Wiley and Sons, New York, 1957.

On a higher analytic level is

D. Blackwell and A. Girshick, *Theory of Games and Statistical Decisions*, John Wiley and Sons, New York, 1954.

The interested reader may refer to the *Studies in Game Theory* issued by the Princeton University Press.

Finally, let us quote Leibniz:

"So also the games in themselves merit to be studied and if some penetrating mathematician meditated upon them he would find many important results, for man has never shown more ingenuity than in his plays."

(Taken from

O. Ore, "Pascal and the Invention of Probability Theory," *Amer. Math. Monthly*, vol. 67, 1960, pp. 409–416.)

14.8 For a discussion of various computational techniques, see

G. Brown, "Iterative solutions of games of fictitious play," *Activity Analysis of Production and Allocation*, Cowles Commission, Monograph 13, New York, 1951.

R. Bellman, *On a New Iterative Algorithm for Finding the Solutions of Games and Linear Programming Problems*, The RAND Corporation, Paper P-473, 1951.

R. B. Braithwaite, "A terminating iterative algorithm for solving certain games . . . ," *Naval Research Logistics Q.*, vol. 6, 1959, pp. 63–74.

These methods have been greatly perfected in recent work by J. Kemeny and collaborators.

The Brown-von Neumann technique is particularly interesting since it shows that the value of the game can be obtained as the steady-state behavior of a dynamic process. This approach can be used to attack a variety of variational processes; see, for example,

A. Boldyreff, "Determination of the maximal steady state flow of traffic through a railroad network," *J. Oper. Res. Soc. Amer.*, vol. 3, 1955, pp. 443–465.

Here the author discusses his "flooding" technique, a method closely related to the relaxation methods of Southwell.

The methods of linear programming can also be used to compute the value of a game.

14.9 Various versions of two-handed poker are discussed in the book by von Neumann and Morgenstern cited above. Multistage versions are considered by

R. Bellman and D. Blackwell, "Some two-person games involving bluffing," *Proc. Nat. Acad. Sci. USA*, vol. 35, 1949, pp. 600–605.

191

R. Bellman and D. Blackwell, "Red Dog, Blackjack, and Poker," *Scientific American*, vol. 184, 1951, pp. 44–47.

R. Bellman, "On games involving bluffing," *Rend. del Cir. Mate. di Palermo*, vol. 1, 1952, pp. 1–18.

Other games are considered in the *Studies in Game Theory* cited above, including some three-person games using the Nash theory of the independent point.

14.10 The study of multistage games was initiated in the form of "games of survival,"

R. Bellman and J. P. LaSalle, *On Non-zero-sum Games and Stochastic Processes*, The RAND Corporation, Research Memorandum RM-212, August 19, 1949.

R. Bellman and D. Blackwell, *On a Particular Non-zero-sum Game*, The RAND Corporation, Research Memorandum RM-250, Sept. 27, 1949.

The subject has been intensively investigated since; see the various *Studies in Game Theory* cited above, and

R. Bellman, *Dynamic Programming*, Princeton Univ. Press, Princeton, New Jersey, 1957,

where further references may be found.

The study of games of survival of continuous type leads to functional equations of the form

$$f_T = \min_v \left[g(c, v) + h(c, v) f_{cc} \right],$$

as well as more complex versions involving min-max operators.

14.12 The subject of continuous pursuit games has been intensively investigated by R. Isaacs, who resolved a number of special games, and developed a general theory of this class of problems; see

R. Isaacs, *Games of Pursuit*, The RAND Corporation, Paper P-257, Nov. 17, 1951.

————, *The Problem of Aiming and Evasion*, The RAND Corporation, Paper P-642, March 14, 1955.

————, *Differential Games—I, II, III, IV*, The RAND Corporation, Research Memoranda RM-1391, RM-1399, RM-1411, RM-1486, 1954.

The subject is a good deal more difficult, and nonintuitive, than might be imagined. See, for example, the result of Besicovitch cited in

J. E. Littlewood, *A Mathematician's Miscellany*, Methuen and Co., London, 1953.

For various other results in the theory of multistage games of continuous type, see

W. Fleming, *On Differential Games with Integral Payoff*, The RAND Corporation, Paper P-717, 1955.

H. Scarf, *On Differential Games with Survival Payoff*, The RAND Corporation Paper P-742, 1956,

where many other references can be found.

Finally, let us point out that in some cases the artificial introduction of min-max processes in place of straight maximization or minimization can lead to a quite simple analytic solution. See

R. Bellman, I. Glicksberg, and O. Gross, "Some nonclassical problems in the calculus of variations," *Proc. Amer. Math. Soc.*, vol. 7, 1956, pp. 87–94.

For an interesting exposition of the theory and history of curves of pursuit, see

C. C. Puckette, "The curve of pursuit," *Math. Gaz.*, vol. 37, 1953, pp. 256–260.

A. Bernhart, "Curves of pursuit—II," *Scripta Math.*, vol. XXIII, 1957, pp. 49–66.

————, "Polygons of pursuit," *Scripta Math.*, vol. XXIV, 1958, pp. 23–50.

For some applications of the theory of multistage games to economic processes, see

M. Shubik and G. Thompson, "Games of economic survival," *Naval Res. Log. Q.*, vol. 6, 1959, pp. 111–123.

M. Shubik, *Strategy and Market Structure*, J. Wiley and Sons, New York, 1958.

CHAPTER XV

ADAPTIVE PROCESSES

Believe nothing, no matter where you read it,
   or who said it,
no matter if I have said it,
   unless it agrees with your own reason and
your own common sense.

BUDDHA

15.1 Introduction

In the foregoing chapters, we have discussed in some detail two types of
processes, those of deterministic nature and those of stochastic nature.
In so doing, we have continually emphasized the many assumptions that
went into the formulation of these processes as mathematical images of
physical processes, as well as the many analytic and computational
difficulties attendant upon their solution. Despite the fact that we introduced some novel techniques to treat the associated control processes, we
adhered rigidly to a classical conceptual basis.

In this part of the book, we wish to enter uncharted and more rugged
territory. A more precise analogy would be that we are penetrating domains
of mathematics which were not even thought of twenty-five years ago, and
therefore not even blank regions on the map.

It is certainly not unexpected that these more formidable problems will
give rise to mathematical questions of greater difficulty. What is surprising
is that the very construction of a suitable mathematical framework to
house these problems, and the choice of appropriate mathematical tools
to resolve them, are matters which now require considerable care and a
certain amount of ingenuity.

The mathematician is by now accustomed to intractable equations, and
even to unsolved problems, in many parts of his discipline. However, it is
still a matter of some fascination to realize that there are parts of mathematics where the very construction of a precise mathematical statement
of a verbal problem is itself a problem of major difficulty.

As we shall see, what is required for our further studies is a hierarchy
of models of uncertainty of increasing level of sophistication. Each of these
must be specifically tailored to fit the needs of the many new types of

multistage decision processes. Some of these are new in the sense that they finally have been recognized as important, and others which have just arisen from new developments in various scientific fields are new in all senses of the word.

It seems rather clear from even a preliminary survey of this new field of research upon which we are embarking that no single formulation, and no single method, will be powerful enough or comprehensive enough, to treat the many different types of questions that can be asked, or to furnish the many different types of answers that are required. What will be necessary will be a combination of many different ideas and techniques, skillfully blended.

In what follows, we wish to show what can be done using the theory of dynamic programming. Our aim is to show how functional equations and the principle of optimality furnish uniform techniques which permit us to treat a wide variety of questions in a uniform way. Let us emphasize, at the very outset, that it would be extremely dangerous to restrict ourselves in our wider studies to any single theory of adaptive processes. The great challenge in the study of these novel processes is not so much how to solve a particular problem by means of a particular method, as how to find new methods which can be used to furnish solutions to whole classes of problems.

15.2 Uncertainty Revisited

In the previous pages we have considered physical systems whose states at any time were specified by state vectors, which we took to be finite dimensional. In addition, the decision processes we studied involved transformations which changed an initial state into either a precise terminal state, or into a precise distribution of terminal states.

We wish now to discuss what can be done when some of this information, which would appear to be essential for any mathematical study, is lacking. These new processes are treated not only because of the many intriguing mathematical problems which spring from this magic ground at the tap of the mathematician's pencil, but also because of the throng of questions of this nature which the outside world insistently thrusts upon us.

Let us now discuss in a rather casual and intuitive way how problems of this nature arise in a number of unrelated fields.

Given a set of objects with certain probabilistic qualities, the mathematical theory of statistics tells us what tests should be made, and how the results of these tests should be interpreted, in order to arrive at certain desired conclusions as quickly as possible.

Classically, a testing program involved a sequence of tests, at the end of which analysis of the results led to a decision. This type of testing policy corresponds to the classical view of the calculus of variations, discussed in previous chapters, in the sense that the whole testing program is regarded as a single complex operation. In recent times, it was recognized that it might be quite profitable to use the results of preceding tests to determine the course of the testing program itself. In other words, we wish to regard the testing operation as a sequence of decisions. This idea was systematically developed by Wald, leading to the modern theory of sequential analysis.

In these approaches, it is assumed that the distribution of the underlying stochastic quantities is known. A question of great interest and importance is that of developing a corresponding theory to treat the cases where the initial distributions are not completely known, but must be determined on the basis of the tests that are made.

Many of these ideas of modern statistics have been applied with great success to problems of signal detection which fall under the heading of *prediction theory, communication theory*, and *information theory*. The basic problem is that of transmitting a signal which we know in advance is going to be contaminated by disturbances which we call "noise." One aspect of the problem is that of separating the original signal from the received signal, while another aspect is that of transmitting the original signal in such a way that this separation can be performed either perfectly, or with some guaranteed accuracy.

A great deal of effort has been devoted to these problems under the assumption that the disturbances corresponded to random variables with known distributions. A problem of great significance is that of modifying the procedures obtained to date in such a way as to take account of the fact that in many cases these distributions are not known initially. These are questions from the electronics field.

In the area of mathematical economics and operations research, we meet a class of problems which are known under the name of *"optimal inventory."* The general problem is that of stocking a supply unit with a sufficient number of items from time to time so as to be able to meet a stochastic demand for these items from time to time. Although the problems are of some difficulty, a great deal can be done using various techniques, provided that we assume that we know the distribution of the demand over time, and the cost involved in stockpiling, and the costs due to over supply and under-supply. Relatively little has been done towards the more realistic study in which the distribution of demand, and perhaps some of the other data, must be obtained in the course of the process.

The field of medical diagnosis presents us with a number of interesting and significant classes of processes in which certain fundamental data can only be obtained in the course of experimentation, which ideally should be guided by a knowledge of these facts. Here we encounter processes in which the symptoms determine the treatment, and the treatment determines the subsequent treatment. Given the not too rare combination of an opinionated specialist such as one addicted to the psychosomatic school, or to any theory fashionable at the moment, and a patient with a case which is off the beaten path, the results are often tragic. And what is saddest is that the power of preconception is such that not even unpleasant statistics can alter this faith in a favorite theory. The story of Semmelweiss and childbed fever illustrates this very well.

In the field of psychology, we encounter *"learning"* processes, which are specifically aimed at the problem of understanding how people learn new skills and act under uncertainty. Here, attention has so far been focussed upon quite special types of problems, at least as far as any mathematical analysis is concerned. Furthermore, the emphasis has been upon a descriptive theory, rather than a theory devoted to control.

Finally, let us point out that theories of control processes of novel types are necessary for the understanding and construction of robots and automata, of chess-playing machines, of machines which solve geometry problems and logical puzzles, and so on. A "thinking machine" which solves problems by sheer enumeration of cases is easily constructed in many different ways. But no machine which is intended to solve problems of any significance can operate solely on this principle.

The simplest problems in many fields of interest often involve numbers of possibilities of the order of 10^{100}. To treat questions of this type we must think along quite different lines. What are these? Is it reasoning by analogy? Is it use of approximate policies with feedback correction? Nobody knows, and nobody knows how the human brain performs the miracles it does.

At the moment we are groping for clues, and are quite humble. The true research man in these fields makes no extravagant claims as to what he can do now, or as to what can be done ten or one hundred years from now. Furthermore, there is no guarantee that we shall ever possess an adequate theory of human knowledge, or even one much better than we possess now.

This is not a metaphysical remark, but is based upon the following facts. What defeats us at the present time in much of our work in many fields is the "curse of dimensionality," discussed in some detail in Chapter V. We can handle functions of a few variables with some aplomb and view sets of quantities totalling 10^6 or 10^7 with sangfroid.

As soon as we begin thinking in terms of functions of a hundred or a thousand variables, and of sets containing 10^{50} or 10^{1000} elements, we run into problems whose very consideration is beyond our understanding, unless these problems are of quite special type.

Let us now combine this observation with the fact that current research indicates that the smallest physical entities and the smallest biological entities possess fantastically complex structures. It may well be that each element of these structures possesses an equally complex substructure.

To quote De Morgan,

"Great fleas have little fleas upon their backs to bite them, and little fleas have lesser fleas, and so ad infinitum."

Considering the complexity that even one additional independent variable adds to analytic and computational solutions, it is quite possible to envisage a situation in which experiment will uncover more and more phenomena at a rate which swiftly will outstrip the theorist. To some extent this is the situation in nuclear physics today and there is no reason why it should not be the situation in neurology tomorrow.

We are the end products of an evolution lasting hundreds of millions of years with billions of mistakes and dead ends. That one can by the force of analysis and the power of imagination completely unravel this tangled skein is not probable. Since it is the ultimate challenge, perhaps it is just as well.

It is clear, from the foregoing brief description and discussion, that the type of problem which we wish to consider is one of great scientific importance, which cuts across a number of fields. Our aim will be to show how we can formulate and treat a number of processes of this nature by means of the functional equation methods we have so far employed for deterministic and stochastic processes.

As usual, having voiced our qualms and caveats, we will now devote ourselves single-mindedly to the mathematical aspects.

15.3 Reprise

In order to set the stage for our subsequent discussion, and to motivate much of it, let us briefly review the various concepts we have introduced, together with the mathematical structures with which we surrounded them.

We began with a model of a multistage decision process in which it was blithely assumed that:

1. The state of the system is known at each stage of the process, prior to the decision to be made at that stage.

2. The set of possible decisions is known at each stage.

3. The effect of a choice of any member of this set is known at each stage.

4. The duration of the process is known in advance.

5. The criterion function is prescribed in advance. Processes of this type we named *deterministic*.

Upon this framework, but no longer as blithely, we superimposed various elements of uncertainty, by means of the joint concepts of stochastic variables and probability distributions. Agreeing to evaluate the outcome of a decision process by means of expected values, we can enlarge the applicability of the foregoing model by allowing:

1. An unknown initial state, with given probability distribution.

2. A distribution of allowable sets of decisions at each stage of the process.

3. A distribution for the outcomes of any particular decision.

4. A distribution for the duration of the process, or, equivalently, a probability of the termination of the process at any stage, dependent upon the state and the decision made.

5. A distribution for the criterion function to be used to evaluate the sequence of decisions and states.

These distributions can be fixed in time, or taken to depend upon the history of the process. This dependence, however, is taken to be known in advance.

Processes of this type we called *stochastic processes*. One of the principal aims of the preceding chapters was the demonstration that both deterministic and stochastic control processes could be treated by means of the same analytic technique, functional equations. Regardless of the enormous gulf in ideology, a common mathematical approach unifies these two diverse types of processes.

To eliminate the rather troublesome assumption that the probability distributions were known, we invented a second decision-maker called "nature." The sole purpose of the deus ex machina was to choose at each stage of the process the probability distribution most embarrassing to us.

This "dog in the manger" attitude, although interesting and useful, suffers from conservatism. Clearly, in general, it would cost us far too much to pursue policies recommended on this basis.

We now wish to weaken our assumptions in various ways so as to make maximum use of the sequential nature of the processes we are considering. As usual, the idealism which demands an exact formulation must, to some extent, yield to the pragmatism which aims at numerical results.

15.4 Unknown—I

Let us now discuss briefly various types of uncertainty which must all be classified under the general heading of *unknown*.

To being with, we can suppose that the system with which we are dealing has definite, well determined properties, which are not known to us. Thus, a patient may or may not have cancer, a coin may or may not have two heads.

In some cases, these properties will not change in time; in other cases, they may change in a known way, dependent upon or independent of any decisions that we may make.

From here it is a simple step to the study of systems in unknown states which change in deterministic, but unknown, fashion as time goes on. A particular case of this is that where the decisions we make have definite, but unknown effects upon the system.

The basic premise here is that the process is deterministic, but that we do not understand the mechanism. A particular case, of some interest, is that where the system can be one of a set of known systems.

We may also consider situations in which the system possesses hidden state variables, which must be discovered as the process continues.

Related questions arise in situations in which all the requisite information is available, but where we do not have the time to use it all.

15.5 Unknown—II

Since a mathematical model of the foregoing type presents many difficulties, we generally shift to a stochastic mechanism. Here, it is the probability distribution that is considered to be unknown.

The problem can be left in this form, or we can narrow it by supposing that the probability distribution is a member of a particular class of distributions characterized by a small set of parameters. For example, the distributions may all be Gaussian with unknown mean and variance.

We can now begin all over again at this level. These parameters may be taken as fixed but unknown, or as stochastic with unknown probability distributions.

It is clear that continuing in this fashion, we can formulate a hierarchy of processes, each containing the following levels, but each level more complex than all those which have preceded.

15.6 Unknown—III

This partial enumeration of cases by no means exhausts the fascinating possibilities implied by the innocuous term "unknown."

We may not, for example, know initially whether the process under study is deterministic or stochastic, or even whether we are dealing with a process in which we alone are the only decision-makers, or whether decisions are also being made by forces wholly or partially opposed to our interests. This is related to the concept of "hidden variables."

In the following chapters we shall discuss only one type of process described above. The techniques we employ can be applied, however, to treat all of the other types. As stated above, our aim is to show how dynamic programming affords a unified treatment. We certainly do not wish to imply that it is necessarily the only such treatment, or even the most reasonable, or the most efficient in any particular instance.

The danger of subscribing too wholeheartedly to any particular type of mathematical model and to any particular type of mathematical analysis cannot be stressed too emphatically. In dealing with any realistic process where mathematical analysis seems profitable, before plunging into a sea of equations and before being engulfed by calculations, it is essential to examine most carefully the assumptions that are being made and the results that one expects to obtain.

Quite often, a simpler mathematical model based upon sounder premises will yield far more than a quite sophisticated mathematical model based upon a less firm foundation. Approximations, after all, may be made in two places—in the construction of the model and in the solution of the associated equations. It is not at all clear which yields a more judicious approximation.

At the present time, although we possess a comprehensive theory of approximation to functions and to the solutions of equations, we have no scale for the comparison of different mathematical models of a physical process. This is a subject that needs some concentrated effort.

15.7 Adaptive Processes

When faced with uncertainties of the various kinds described above, the common sense approach is to learn from experience. As the process unfolds, we should be able to learn more and more about the unknown structures and unknown parameters.

As mentioned above, the name *learning process* has been attached to processes of this nature which arise in psychological investigations. In order to avoid over-entanglement with one particular field, we shall call this type of process, wherever it occurs, an *adaptive process*.

There is no particular point in trying to give a precise definition of processes of this nature, since each manifestation, in engineering, economics, biology, computers, and so on, will possess individual features of interest.

Processes of this type which we encounter in the study of feedback control processes will be called *adaptive control processes*.

Bibliography and Comments

15.1 For a general discussion of these and other problems, see

W. Ross Ashby, *Introduction to Cybernetics*, John Wiley and Sons, New York, 1957.

N. Wiener, *Cybernetics*, John Wiley and Sons, New York, 1948.

R. R. Bush and F. Mosteller, *Stochastic Models for Learning*, John Wiley and Sons, New York, 1955.

A. M. Uttley, "A theory of the mechanism of learning based on the computation of conditional probabilities," *Proc. First International Congress on Cybernetics*, Namur, 1956, pp. 830–856.

——————, "Conditional Probability Computing in a Nervous System," *Proc. Sym. on the Mechanization of Thought Processes*, National Phys. Lab., England, 1958.

J. von Neumann, *The Computer and the Brain*, Yale Univ. Press, New Haven, 1956.

F. Rosenblatt, *The Perceptron—A Theory of Statistical Separability in Cognitive Systems*, Cornell Aero. Lab. Inc., Report Nos. VG 1196, G-1, 1958, 1196, G-2, 1958.

15.2 The reader interested in the approach to decision processes by way of mathematical statistics may wish to refer to

A. Wald, *Sequential Analysis*, John Wiley and Sons, New York, 1947.

A. Wald, *Decision Processes*, John Wiley and Sons, New York,

D. Blackwell and A. Girshick, *Theory of Games and Statistical Decisions*, John Wiley and Sons, New York, 1954.

For an extensive discussion of signal detection, see

D. Middleton, *Random Processes, Signals and Noise—An Introduction to Statistical Communication Theory*, McGraw-Hill Book Co., Inc., New York, 1960.

For a treatment of the inventory problem by means of classical variational tools, and by means of dynamic programming techniques, see

K. D. Arrow, S. Karlin, and H. Scarf, *Studies in the Mathematical Theory of Inventory and Production*, Stanford Univ. Press, Stanford, Calif., 1958.

R. Bellman, *Dynamic Programming*, Princeton Univ. Press, Princeton, N.J., 1957,

and the work by Scarf, Dvoretzky, Kiefer, and Wolfowitz referred to at the end of Chapter XVI.

For the first mathematical treatment of an adaptive process occurring in the medical field, see the papers by W. R. Thompson cited at the end of Chapter XVI. These papers were the origin of the "two-armed bandit" problem which has attracted a good deal of attention in recent times, and are pioneering papers in the field of sequential analysis. The two-armed bandit problem is discussed in Chapter XVI.

For a treatment of communication processes along adaptive lines, see Chapter XVII. A number of references to the extensive work by M. Flood in the theory of adaptive processes will be found at the end of Chapter XVI, along with many other references to closely related work by others.

For a discussion of the neurological aspects of memory, see

W. Penfield, "The interpretative cortex," *Science*, vol. 129, 1959, pp. 1719–1725.

W. Penfield and L. Roberts, *Speech and Brain-Mechanisms*, Princeton Univ. Press, Princeton, N.J., 1959.

We wish to thank the distinguished sculptor and neurosurgeon, Emil Seletz, for many informative and entertaining discussions on matters of this nature.

CHAPTER XVI

ADAPTIVE CONTROL PROCESSES

> And as imagination bodies forth
> The form of things unknown, the poet's pen
> Turns them to shapes, and gives to airy nothing
> A local habitation and a name.
>
> WILLIAM SHAKESPEARE
> *"A Midsummer Night's Dream,"*
> Act V, Scene I.

16.1 Introduction

We have indicated in the previous chapter of expository nature how it is that in many processes of economic, engineering, biological, psychological, and statistical origin a decision-making device is required to perform under various conditions of uncertainty regarding the intrinsic physical process. This decision-making device may be human, inanimate, or a combination of both elements. Problems which are difficult enough, as we have discussed in the sections on deterministic control processes, under ideal conditions where complete information is presupposed, become truly formidable in their more realistic setting.

The available information ranges from the approximation of complete knowledge on one hand to the approximation of complete ignorance on the other. What prevents our analysis of these new problems from being rather sterile is the fact that as the process unfolds, additional information becomes available to the control device. This additional data is either accumulated automatically, or as a result of deliberate probing which itself is part of the control policy. It follows that there exists the possibility of having the control device *"learn,"* which is to say, improve its performance on the basis of experience. Here, experience is used to denote a suitable combination of observation and analysis.

It is rather important to stress the fact that there is nothing mystical or vague about this type of learning, provided that we are talking in terms of digital and analogue computers. The way in which the machine will learn in any particular process is part of the program that regulates the operation of the computer from the very outset of the process. Machines of this type can be called "thinking machines," if one wishes to indulge one's fancy, but, as we have already commented in rather strong terms, any attempt to detect any close resemblance between the behavior of these devices and the behavior of the human brain is definitely premature on the basis of our current understanding.

When we talk about learning in terms of the human operator, then we enter a field in which concepts and theories, results and experiments, alike are vague and ill-formed. How the human brain operates in storing and utilizing knowledge is a mystery far beyond the present level of neurological research.

At the present, the hope is not so much to study computers in order to learn something about the brain, but rather to study the brain so that we can learn how to construct really efficient computers. If we could combine the memory and reasoning capacity of the human brain with the accuracy of the computer, then we would have an instrument with which to tackle some of the fundamental problems of the universe.

As it is, we must content ourselves first with constructing a broad mathematical foundation for the treatment of a number of classes of significant adaptive processes, and secondly with using this to obtain the numerical solution of a number of simplified mathematical models. Parenthetically, let us remark that it is often impossible to assess the conceptual and analytical level of difficulty of a problem from its verbal formulation. Many questions easily posed in a few words from everyday usage escape even a formulation in precise terms.

Using the theory of dynamic programming in a fashion similar to that employed in the study of deterministic and stochastic processes, we shall first present a mathematical formulation of adaptive control processes in quite abstract terms. Then, step-by-step, we shall show how to reduce this level to the point where we have constructive algorithms for the solution of numerical problems.

We shall illustrate the actual mechanism by means of two examples. The first, a feedback control problem, will be presented in deterministic, stochastic, and adaptive form, in order to illustrate the similarities and differences that exist in the respective treatments. The second problem which we shall discuss is commonly called the "two-armed bandit" problem, a transliteration of a question which first arose in 1935 in the use of "wonder drugs."

In the foregoing chapter, we aired our doubts concerning the inadequacies of the mathematical model we shall employ along with various other philosophical trepidations. In this chapter, we retreat from the world of reality into the world of mathematics. The processes that we consider are now mathematical processes, precisely formulated on an axiomatic basis, with a precise definition of an optimal policy. The question of the relevance of the solution of the mathematical problem to the solution of the physical problem is for the moment put aside.

16.2 Information Pattern

We now wish to introduce a new concept, one which will be basic to our further work in this field. This is the idea of an *information pattern*. Like many other fundamental concepts, it appears quite futile to attempt to define it precisely. Any attempt to do so results inevitably in the introduction of many other ideas of far less intuitive nature and of a greater or equal level of sophistication.

At any time, the state of the system will be specified by a point p in phase space, as usual, and by an information pattern P. This information pattern represents the information about the past history of the process which we wish to retain in order to guide our further actions. The best way to render this idea familiar is to give some examples.

Suppose that we are engaged in a coin-tossing process with a coin which we have never seen before. At each stage of the process, keeping in mind the past record of heads and tails, we are required to predict whether the coin will fall heads or tails on the next toss.

The information pattern could be the actual sequence of heads and tails that has been observed over the previous N tosses, a sequence of the form

(16.1) $$HHTTT \ldots HT,$$

where H stands for heads and T for tails, or it could merely be the total number of heads that have been observed and the total number of tails that have occurred. The point is there is no rigid rule which determines what an information pattern must be. It is up to us to choose it.

In the sports world, these information patterns are of extreme value and very often make the difference between a victory and a defeat. On the baseball field, when a new batter comes into the league, the pitchers test him on many different types of pitches. They try fast balls, change-ups, inside and outside pitches, curves and above all, "dusters." Many a potential home-run king has gone back to the minors because of an inability to hit an inside curve, or because of an inability to stand up to a "duster." A good catcher keeps "book" on each batter and tries both to have the pitcher pitch to his weakness, and to mix up the pitches. Shades of game theory!

The important point to stress for our purposes is that the information pattern is the past history of the batter's successes and failures against different types of pitches and pitchers. For a new batter, this information pattern changes from month to month, and from year to year.

Similarly, in the football sphere, pictures of every game are run over and over in slow motion to watch the behavior of the opposing quarterback when calling signals. Does he wet the fingers of his right hand before throwing a pass? Does he always glance towards the right side of the line when planning a left-end run? Or, are these gestures observed in only three or four games purely coincidental?

This is all part of the information pattern that accumulates concerning a team and its individual players. The decisions made by the coach and the players on one team depend upon the actual state of the game, and upon the data they have accumulated.

Yardley, in his already classic book on Poker, tells of such a learning process which enabled him to determine when a particular player was bluffing and when not.

Let us now turn to the economic world. When a new machine is introduced, a number of questions arise concerning the quantity and quality of servicing equipment and spare parts. Although some estimates can be gained a priori from preliminary tests, in general, precise information is only available after the device has been in operation for a while. In this situation, preliminary estimates must constantly be revised on the basis of the actual demands for spare parts, servicing, and so on, that are made from time period to time period. This is an important aspect of *"optimal inventory"* problems that has as yet been little investigated.

As a final example, let us turn to the scientific world. When a physicist or mathematician sits down to construct a mathematical model of a physical process, he has in his memory the results of a number of previous attempts of the same type. Some of these have been outstandingly successful, some have been moderately so, and some have failed completely. In addition to this, the scientist possesses a number of partial and complete explanations and rationalizations for the success and failure of these previous attempts. Finally, he is guided by some metaphysical principles, such as determinism or indeterminism, and by certain physical and mathematical esthetics. All of this constitutes his information pattern.

As a result of a number of consecutive failures or successes, this information pattern changes in several ways. Not only is the direct count of successes and failures affected, but far more important, the belief in basic principles may also be affected, and eventually result in a complete rejection of theories of one type or another.

We have seen examples of this in the rise of quantum theory and relativity theory. The existence and success of these theories should be very amusing and instructive to the psychologist and psychiatrist, since we have here very clear examples of scientists professing and using principles which they on the whole resolutely do not believe in, both psychologically and emotionally.

The problems involved in choosing useful information patterns and in expressing these in reasonable analytic form are of great subtlety and difficulty. Usually, a certain amount of both experience and ingenuity is required. Some examples will be discussed below.

16.3 Basic Assumptions

Logic, observation, and frustration, in what proportions is not at all clear, have forced us successively from our citadel of determinism to the, by now, fairly civilized fields of stochastic processes, and finally to the jungle of adaptive processes.

In order to proceed in the study of these processes, we must make some assumptions. These will be the following.

A. (i) The state of the process is specified at any stage by a point p in phase space and by an information pattern P.

   (ii) The set of available decisions, or transformations, is determined at any stage by this combination, (p, P).

  (iii) As a result of a decision q, the "point" (p, P) is transformed a priori into a point (p_1, P_1), a stochastic quantity whose distribution is determined by the initial point (p, P).

  (iv) After the decision has been made, the a posteriori point, (p_2, P_2), is observable.

   (v) The criterion function governing the process at any stage is determined by (p, P).

In this general fashion, we wish to cover a number of situations that can arise in practice. In the first place, the actual state of the system may not be known precisely. Some of the components may be known, others may possess precise distribution functions, and still others may only be known in the sense that some past set of values assumed by these parameters is known. This is the substance of the assumption in A(i).

Secondly, in many situations, we may not initially be aware of all the possible decisions that can be made. As the process continues, and as a result of our actions, the set of available decisions changes. This is quite typical of various experimental processes.

The third assumption is familiar to us from the study of stochastic control processes. What is different here is that even after the decision has been made, the new state variable need not be a point in some finite dimensional space, but may still remain a combination of the form (p_1, P_1).

The substance of the fourth assumption is that even though this new point, (p_1, P_1), is observable, we may still not wish to expend the resources required to observe it, and may prefer to continue purely on the basis of the a priori information.

Finally, the fifth assumption is designed to cover processes in which the ultimate purpose of the process is not clear initially. As the process continues, we revise our objectives on the basis of what actually occurs.

A particular case of this is that where the duration of the process is not precisely known initially.

Since our aim in this volume is to introduce the reader to the study of these new problems and new techniques, we feel that there is no need at this stage to study the field of adaptive control processes in its full generality. Consequently, in the section that follows we will take advantage of only a few of the new degrees of freedom offered by adaptive processes. It should, however, be clear from the results given below, how one can proceed in the more general situations.

16.4 Mathematical Formulation

Let us consider a system specified by a point p in phase space and the information pattern P. We shall say that the system is specified by the point (p, P). As a result of a decision q, p is transformed into a new point in phase space, p_1, given a priori by the stochastic transformation

$$(16.2) \qquad p_1 = T_1(p, P; q, r),$$

and the information pattern P is transformed into a new information pattern

$$(16.3) \qquad P_1 = T_2(p, P; q, r),$$

where T_2 is another stochastic transformation. The variable r is a stochastic variable, specified by a probability distribution, $dG(p, P; q, r)$, itself a part of the information pattern P.

We shall suppose that the transformations T_1 and T_2 are known, although in many adaptive processes, the determination of these functions, and, as mentioned above, even of the information pattern itself, is an essential part of the general problem associated with the control process.

With this format, let it be required to determine a sequence, $[q_1, q_2, \ldots, q_N]$, which will minimize the expected value of a preassigned function of the terminal state, $[p_N, P_N]$. The expected value is taken with respect to the set of a priori probability distributions.

16.5 Functional Equations

Following our usual approach, we introduce the sequence of functions

$$(16.4) \qquad f_N(p, P) = \text{Min Exp } \phi(p_N, P_N),$$

where ϕ is the preassigned function mentioned above. Then, as before, the principle of optimality yields the recurrence relation

$$(16.5)$$
$$f_N(p, P) = \underset{q_1}{\text{Min}} \left[\int f_{N-1}[T_1(p, P; q, r), T_2(p, P; q, r)] \, dG(p, P; q, r) \right],$$

208

$N = 2, 3, \ldots$, with

(16.6)
$$f_1(p, P) = \operatorname*{Min}_{q_1} \left[\int \phi[T_1(p, P; q, r), T_2(p, P; q, r)] \, dG(p, P; q, r) \right].$$

These relations can be used to establish the existence of optimal policies, various structural characteristics of the solution, and so on. At the moment, however, we are following the spoor of other game.

16.6 From Information Patterns to Distribution Functions

As a first step towards the clarification of this perhaps distressingly vague formulation, let us replace the amorphous information pattern by a distribution function.

Let us consider a situation in which we know the state of the system at any stage, we know the set of allowable decisions, and the criterion function, but where we do not know the transformation which results from a decision q.

We shall suppose that we possess an a priori distribution function $dG(z, p, q)$ for the stochastic outcome z of a decision q made in state p, and that we possess a procedure for modifying this a priori estimate on the basis of observation of the actual state resulting from the decision q, namely p_1, and the information we already possess, namely p_0, q and $dG(z, p, q)$.

Thus, as a result of the decision q_1,

(16.7)
$$p_0 \rightarrow p_1$$
$$dG(z, p, q) \rightarrow dH(z, p, q; p_0, G, q_1, p_1).$$

This notation serves to indicate the fact that the new a priori estimates depend upon the old estimates, the former state, the new state, and the decision made. What form this transformation takes in specific cases will be discussed below. The system is then in the new physical state, p_1, and possesses the new *information pattern*, dH. Observe that here the information pattern is a familiar object, a distribution function.

16.7 Feedback Control

Once again to simplify notation, let us consider only a terminal control process. On the basis of the assumptions made in the foregoing section, starting in the "state" $[p, G(z, p, q)]$, we wish to make a sequence of decisions which will maximize the expected value of a prescribed function, $\phi(p_N, G_N)$, of the terminal state. Here the duration of the process is fixed and known.

16.8 Functional Equations

With the foregoing as preparation, there is now no difficulty in writing down functional equations which describe the unfolding of the process. Let

(16.8) $f_N[p; G(z, p, q)] =$ the expected value of $\phi(p_N, G_N)$
obtained using an optimal policy
for an N-stage process, starting
in state (p, G).

The relation of (16.5) now yields the recurrence relation

(16.9)

$$f_N[p; G(z, p, q)] = \underset{q_1}{\text{Max}} \int_w f_{N-1}[w; H(z, p, q; p, G, q_1, w)] \, dG(w, p, q_1),$$

for $N = 2, 3, \ldots$, and we clearly have the result

(16.10) $f_1[p; G(z, p, q)] = \underset{q_1}{\text{Max}} \int_w \phi[w, H(z, p, q; p, G, q_1, w)] \, dG(w, p, q_1),$
for $N = 1$.

As noted above, these equations can be used to derive existence and uniqueness theorems for return functions and optimal policies, and also to obtain certain structural properties of the solution.

When we turn to the analysis of processes which are too complex for a direct analytic approach, as is invariably the case with realistic models, we wish to seek the succor of a computational solution. The occurrence of functions of functions, such as the sequence $f(p, G)$, effectively rules this out. What then do we do?

16.9 Further Structural Assumptions

In order to make some further progress in our reduction process, we must make some further assumptions. To reduce the foregoing recurrence relations to a more manageable form, we shall assume that the basic structure of the distribution function is known. The uncertainty that exists arises only in regard to the value of certain parameters specifying the particular distribution within the fixed family of distributions.

At any stage of the process, in place of an a priori estimate for the distribution function, we shall suppose that we possess an a priori distribution function for certain parameters appearing in the distribution function.

Let us give some examples of families of distribution functions determined by parameters. Probably the most familiar and useful family is that of Gaussian functions,

(16.11) $$dG = e^{-\left(\frac{z-m}{\sigma}\right)^2} \Big/ \sigma\sqrt{\pi}.$$

Here one or both of the parameters, σ and m, vary from stage to stage on the basis of the observed values of the stochastic variables, $z_1, z_2, \ldots$.

Another important example, and one which we shall make use of in the following chapter, is the family associated with a stochastic variable which assumes only two values, say 0 and 1.

Let

$$(16.12) \qquad \begin{aligned} z &= 0, \quad \text{with probability } p \\ &= 1, \quad \text{with probability } q = 1 - p. \end{aligned}$$

This family of distributions can be associated with the tossing of a coin, letting a head correspond to a zero and a tail correspond to a one.

In the first case, we have at each stage as our information pattern an a priori distribution function for σ and m, $dH(\sigma, m)$. In the second case, we have an a priori distribution function for p, $dH(p)$.

16.10 Reduction from Functionals to Functions

We are now ready to take the decisive step of reducing the functional $f_N(p, G)$ to a function.

In order to do this, we wish to formulate the problem so that the transformation from the information pattern $G(z, p, q)$ to the information pattern $H(z, p, q; p, G, q_1, w)$ is one that can be represented by a point transformation. This will be the case if G and H are both members of a family of distribution functions, $K(z, p, q, \alpha)$, where α is a vector parameter, such that

$$(16.13) \qquad G(z, p, q) \equiv K(z, p, q; \alpha),$$

$$H(z, p, q; p, G, q_1, w) \equiv K(z, p, q; \beta),$$

with β a prescribed function of p, q_1, w, and α,

$$(16.14) \qquad \beta = \psi(p, q_1, w, \alpha).$$

Then we may write

$$(16.15) \qquad f_N[p, G(z, p, q)] \equiv f_N(p; \alpha),$$

completely suppressing the functional dependence, and the basic recurrence relation of (16.9) becomes

$$(16.16) \qquad f_N(p; \alpha) = \underset{q_1}{\text{Max}} \int_w f_{N-1}(w; \beta) \, dK(w, p, q_1; \alpha).$$

We now possess a computational algorithm which in certain carefully chosen cases can be carried out with current computing devices.

16.11 An Illustrative Example—Deterministic Version

Let us now show how these ideas may be applied to the study of control processes. In order to emphasize the difference between this new approach and the previous approaches, let us consider in turn the way in which the same process would be treated first as a deterministic process, then as a stochastic process, and finally as an adaptive process.

Consider a scalar recurrence relation of the form

$$(16.17) \qquad u_{n+1} = au_n + v_n, \quad u_0 = c,$$

where u_n is the state variable and v_n is the control variable. The sequence $\{v_n\}$ is to be chosen so as to minimize the function

$$(16.18) \qquad |u_N| + b \sum_{k=0}^{N-1} u_k^2$$

subject to the constraints

$$(16.19) \qquad |v_i| \le r, \quad i = 1, 2, \ldots, N.$$

Introducing the sequence of functions defined by the relation

$$(16.20) \qquad f_N(c) = \underset{\{v_i\}}{\text{Min}} \left[|u_N| + b \sum_{k=0}^{N-1} u_k^2 \right],$$

we obtain the recurrence relation

$$(16.21) \qquad f_N(c) = \underset{|v_0| \le r}{\text{Min}} [b(ac + v_0)^2 + f_{N-1}(ac + v_0)],$$

for $N \ge 2$, and the initial relation

$$(16.22) \qquad f_1(c) = \underset{|v_0| \le r}{\text{Min}} [b(ac + v_0)^2 + |ac + v_0|].$$

We thus obtain a simple computational solution of the problem of determining optimal feedback control.

16.12 Stochastic Version

In place of the recurrence relation of the previous section, let us use a stochastic transformation

$$(16.23) \qquad u_{n+1} = au_n + r_n + v_n, \quad u_0 = c.$$

Here $\{r_n\}$ is a sequence of independent random variables, assuming only the values 1 and 0. The common probability distribution is determined by the relations

$$(16.24) \qquad r_n = 1 \quad \text{with probability } p,$$
$$r_n = 0 \quad \text{with probability } 1 - p.$$

The quantity p is taken to be known, and for simplicity of notation, taken to be independent of n, and independent of the state of the system, although these features can readily be added if desired.

We now wish to minimize the expected value of the quantity appearing in (16.18), a simple example of a stochastic control process. Calling the minimum expected value $f_N(c)$, we obtain the relation

$$(16.25) \quad f_1(c) = \min_{|v_0| \le r} \left[\int_{r_0} [|ac + v_0 + r_0| + b(ac + r_0 + v_0)^2] \, dG(r_0) \right]$$

$$= \min_{|v_0| \le r} [p[|ac + v_0 + 1| + b(ac + v_0 + 1)^2]$$

$$+ (1 - p)[|ac + v_0| + b(ac + v_0)^2]],$$

for $N = 1$, and for general N,

$$(16.26) \quad f_N(c) = \min_{|v_0| \le r} [p[b(ac + v_0 + 1)^2 + f_{N-1}(ac + v_0 + 1)]$$

$$+ (1 - p)[b(ac + v_0)^2 + f_{N-1}(ac + v_0)]].$$

Once again, we have a simple computational scheme for determining optimal feedback control.

16.13 Adaptive Version

Let us now consider an adaptive control process. We shall suppose that we are given the information that the random variables r_n possess distributions of the special type described above in (16.24), but that we do not know the precise value of p.

However, we shall assume that we do possess an a priori distribution for the value of p, $dG(p)$, which we agree to regard as the true probability distribution in the absence of further information, and that we possess a definite procedure for modifying this a priori estimate on the basis of observations that are made as the process continues over time. This is a particular illustration of the technique indicated in 16.10 for reducing functionals to functions.

We shall employ a Bayes approach to modify the a priori distribution. If, over the preceding $m + n$ stages, we observe that the random variables, $\{r_i\}$, have taken the value one m times and the value zero n times, and if further, the a priori distribution at the beginning of these $m + n$ stages was $dG(p)$, we agree to take as our new distribution the function

$$(16.27) \quad dG_{m,n}(p) = p^m(1 - p)^n \, dG(p) \Big/ \int_0^1 p^m(1 - p)^n \, dG(p).$$

Once we have fixed upon an a priori estimate for the probability distribution, $dG(p)$, at the beginning of the process, the a priori distribution at any stage of the decision process is uniquely determined in accordance with the above by the numbers m and n. This simple observation enables us to reduce the information pattern from that of a function $dG_{m,n}(p)$ to that of the two numbers m and n. The result is that the state of the process is completely specified at any stage by the physical state c and the two integers m and n.

213

In this way, the problem is reduced from one requiring functionals to one utilizing functions. Not only is this of essential aid in any analytic studies, but it is absolutely essential for computational purposes.

What is rather interesting to contemplate is that the advent of computing machines has actually forced us to exercise our ingenuity in a way which is quite unexpected. In order to obtain a feasible computational scheme, we must do much more than construct an algorithm which is perfectly suitable for the purposes of formulation, existence and uniqueness, and so on. There is now a great premium on nonconformity, on developing special schemes for special problems. In other words, the computing machine which in many ways is the very symbol of regimentation and uniformity, and indeed poses great threats in this direction, also serves as a stimulus to the type of classical analysis based upon the particular that many of the devotees of the ultra-abstract school would like to banish forever.

Let us then introduce the sequence of functions, $\{f_N(c, m, n)\}$, where $f_N(c, m, n)$ is defined once again as the minimum of the expected value of the quantity in (16.18), starting with the information that m ones and n zeros have occurred, that the system is in state c, and that N stages remain.

For $N = 1$, we obtain the relation

$$(16.28) \quad f_1(c, m, n) = \underset{|v_0| \leq r}{\text{Min}} \left[p_{m,n}[b(ac + v_0 + 1)^2 + |ac + v_0 + 1|] \right.$$
$$\left. + (1 - p_{m,n})[b(ac + v_0)^2 + |ac + v_0|] \right],$$

where $p_{m,n}$ is the expected probability obtained from the probability distribution in (16.27), which is to say,

$$(16.29) \qquad p_{m,n} = \frac{\displaystyle\int_0^1 p^{m+1}(1 - p)^n \, dG(p)}{\displaystyle\int_0^1 p^m(1 - p)^n \, dG(p)} .$$

For $N \geq 2$, we have the recurrence relation

(16.30)
$$f_N(c, m, n) = \underset{|v_0| \leq r}{\text{Min}} \left[p_{m,n}[b(ac + v_0 + 1)^2 + f_{N-1}(ac + v_0 + 1, m + 1, n)] \right.$$
$$\left. + (1 - p_{m,n})[b(ac + v_0)^2 + f_{N-1}(ac + v_0, m, n + 1)] \right].$$

Although we now possess a feasible computational algorithm for determining optimal policies, the details of the computation are no longer trivial. This is due to the occurrence of an infinite sequence of functions $\{f_N(c, m, n)\}$ in which the value of $f_N(m, n)$ depends upon the values of $f_{N-1}(m + 1, n)$ and $f_{N-1}(m, n + 1)$. We face the menace of the expanding grid! Various methods of successive approximations must be employed, which we will not enter into here.

Let us also point out that a number of quite interesting results can be obtained for the special cases where the recurrence relations are linear, as above, and the criterion functions are quadratic. As we shall discuss subsequently, the importance of results of this nature resides in the fact that they can be used as a basis for successive approximations to the solutions of more complex processes.

16.14 Sufficient Statistics

The fact that the history of the process described in the preceding paragraph can be compressed in the indicated fashion, enabling us to describe the process in terms of functions rather than functionals, is a particular instance of the power of the theory of "sufficient statistics."

There are a large and important class of special distributions which have the vital property that use of observations à la Bayes yields a member of the same family of distributions. In a still larger number of cases, this occurs asymptotically (e.g. the central limit theorem), affording another approach to this vital problem of data reduction. Many interesting problems arise in this way, which have been little studied.

16.15 The Two-armed Bandit Problem

Let us now briefly discuss a particular adaptive process that has attracted a great deal of attention. It arose originally in connection with the use of untried drugs in critical medical cases.

Suppose that we face two slot machines, unimaginatively called I and II, with the following properties. When the lever on the first machine is pulled, there is a probability r of receiving a gain of one unit, and a probability $(1 - r)$ of receiving nothing. If the second machine is "played," there is a corresponding probability s of gaining one unit, and a probability $(1 - s)$ of gaining nothing.

Consider the case where s is known, but r is not, and let $dF(r)$ be the a priori probability distribution for r. In place of considering a finite process, consider one in which we discount at a constant rate over time. If then z_n is the stochastic quantity gained on the n-th trial, we wish to maximize the expected value of the variable

(16.31)
$$w = \sum_{n=1}^{\infty} a^n z_n.$$

where $0 < a < 1$.

Let

(16.32) $f_{m,n}$ = the expected return obtained using an optimal policy for an unbounded process after the first machine has had m successes and n failures,

and, in the same fashion as above,

(16.33)
$$dF_{m,n}(r) = \frac{r^m(1 - r)^n \, dF(r)}{\displaystyle\int_0^1 r^m(1 - r)^n \, dF(r)}.$$

Since at each stage we have a choice of trying one machine or the other, it is easy to see that

$$(16.34) \qquad f_{m,n} = \text{Max} \begin{bmatrix} \text{I: } \int_0^1 r \, dF_{m,n}(r)[1 + af_{m+1,n}] \\ + \int_0^1 (1-r) \, dF_{m,n}(r)[af_{m,n+1}], \\ \text{II: } s/(1-a). \end{bmatrix}$$

Once again, the computational determination of the sequence $\{f_{m,n}\}$ requires the use of a method of successive approximations.

Bibliography and Discussion

16.1 For a subject which is still so new that no uniform name or terminology or prescribed domain exists, the study of adaptive control processes possesses an amazing bibliography. An excellent list is given in

P. R. Stromer, "Adaptive or self-optimizing systems—a bibliography," *IRE Trans. on Automatic Control*, vol. AC-4, 1959, pp. 65–68.

Books of great interest, overlapping the subjects we have discussed in many different ways are

W. R. Ashby, *An Introduction to Cybernetics*, John Wiley and Sons, New York, 1956.

R. R. Bush and F. Mosteller, *Stochastic Learning Theory*, John Wiley and Sons, New York, 1957.

J. von Neumann, *The Computer and the Brain*, Yale Univ. Press, New Haven, Conn., 1958.

N. Wiener, *Extrapolation, Interpolation and Smoothing of Stationary Time Series with Engineering Applications*, Technology Press, Cambridge, Mass., 1949.

N. Wiener, *Cybernetics*, John Wiley and Sons, New York, 1948.

Articles of importance are

H. J. Muller, "Evolution by mutation," *Bull. Amer. Math. Soc.*, vol. 64, 1958, pp. 137–160.

L. A. Zadeh and J. R. Ragazzini, "An extension of Wiener's theory of prediction," *J. Appl. Physics*, vol. 21, 1950.

H. Robbins, "Some aspects of the sequential design of experiments," *Bull. Amer. Math. Soc.*, vol. 58, 1952, pp. 527–536.

K. Donaldson, "Docile sequence engine, a new type of model for a learning machine," *Research Applied to Industry*, vol. 12, 1959, pp. 155–157.

Let us also cite the significant new techniques of Box and collaborators.

G. E. P. Box, "Evolutionary operations, a method for increasing industrial productivity," *Appl. Stat.*, vol. 6, 1957, pp. 3–23.

G. E. P. Box and J. S. Hunter, "Condensed calculations for evolutionary operation programs," *Technometrics*, vol. 1, 1959, pp. 77–95.

G. E. P. Box and G. A. Coutie, "Application of digital computers in the exploration of functional relationships," *Proc. Int. Elec. Engrs.*, B 103, 1957, supplement no. 1, pp. 100–107,

and the equally significant work in stochastic approximation,

H. Robbins and S. Monro, "A stochastic approximation model," *Ann. Math. Stat.*, vol. 22, 1951, pp. 400–407.

T. Kitagawa, "Successive process of statistical controls," (1), *Mem. Fac. Sci.*, Kyushu Univ., Ser. A., 7 (1952), pp. 13–28.

————, "Successive processes of statistical controls," (2), *Mem. Fac. Sci.*, Kyushu Univ., Ser. A., 13 (1959), pp. 1–16.

A. Dvoretzky, "On stochastic approximation," *Proc. Third Berkeley Symposium on Math. Stat. and Prob.*, vol. 1, 1956, pp. 39–56, Univ. Calif. Press, Berkeley, Calif.

Many additional references will be found in this paper.

J. E. Bertram, "Control by stochastic adjustment," *Applications and Industry*, No. 46, 1960, pp. 485–490.

Let us also mention the work on sequential search processes:

J. Kiefer, "Optimum sequential search and approximation methods under minimum regularity assumptions," *J. Soc. Ind. Appl. Math.*, vol. 5, 1957, pp. 105–136.

S. Johnson, "Optimal search is Fibonaccian," unpublished (the results are given in *Dynamic Programming*, pp. 34–36).

O. Gross and S. Johnson, "Sequential minimax search for a zero of a convex function," *Math. Tables and other Aids to Computation*, vol. XIII, 1959, pp. 44–51.

Interesting expositions are

B. McMillan, "Where do we stand," *IRE Trans. on Information Theory*, vol. IT-3, 1957, pp. 173–174.

L. Zadeh, "What is optimal," *IRE Trans. on Information Theory*, vol. IT-4, 1958, p. 3.

A very interesting and pertinent discussion will be found in

O. Helmer and N. Rescher, *On the Epistemology of the Inexact Sciences*, The RAND Corporation, Report R-353, 1960.

16.2–16.5 We follow the paper

R. Bellman and R. Kalaba, "A mathematical theory of adaptive control processes," *Proc. Nat. Acad. Sci. USA*, vol. 45, 1959, pp. 1288–1290.

H. O. Yardley, *The Education of a Poker Player*, Simon and Schuster, 1957.

16.6–16.10 We follow the discussion in

R. Bellman and R. Kalaba, "Dynamic programming and adaptive processes —I: mathematical foundation," *IRE Trans.*, PGAC, vol. AC-5, 1960, pp. 5–10.

16.11–16.13 We follow

R. Bellman and R. Kalaba, "On adaptive control processes," *1959 IRE National Convention Record*, *IRE Trans.*, PGAC, vol. AC-4, 1959, pp. 1–9.

16.14 An introduction to the topic of "sufficient statistics" may be found in

A. M. Mood, *Introduction to the Theory of Statistics*, McGraw-Hill Book Co., Inc., New York, 1950.

An extensive treatment of linear recurrence relations and quadratic criteria via sufficient statistics will be found in a forthcoming thesis by M. Freimer in the Department of Mathematics at Harvard.

See also

M. Freimer, *A Dynamic Programming Approach to Adaptive Control Processes*, Lincoln Laboratory, 1959; appearing as

M. Freimer, "A dynamic programming approach to adaptive control processes," *IRE Trans.*, vol. AC-4, 1959, pp. 10–15.

For a discussion of adaptive processes with time lags, see

J. D. R. Kramer, Jr., *On Control of Linear Systems with Time Lags*, The RAND Corporation, Paper P-1948, March, 1960.

Computational aspects of adaptive processes will be discussed in a forthcoming thesis by M. Aoki in the Department of Engineering at UCLA, and

M. Aoki, *Dynamic Programming Approach to a Final-Value Control System with a Random Variable Having an Unknown Distribution*, Tech. Report 60-15, 60-16, UCLA, Dept. of Engineering, 1960.

—————, *On Optimal and Suboptimal Policies in the Choice of Control Forces for Final-Value Systems*, IRE International Convention, Session No. 1, March, 1960.

16.15 The original work of W. R. Thompson, unappreciated at the time, may be found in

W. R. Thompson, "On the likelihood that one unknown probability exceeds another in view of the evidence of two samples," *Biometrika*, vol. 25, 1933, pp. 285–294.

—————, "On the theory of apportionment," *Amer. J. Math.*, vol. 57, 1935.

The treatment given here first appeared in

R. Bellman, "A problem in the sequential design of experiments," *Sankhya*, vol. 16, 1956, pp. 221–229.

A discussion using different methods may be found in

S. Karlin, R. Bradt, and S. Johnson, "On sequential designs for maximizing the sum of n observations," *Ann. Math. Stat.*, vol. 27, 1956, pp. 1061–1074.

Papers treating problems of this nature in a different way are

H. Robbins, "A sequential decision problem with a finite memory," *Proc. Nat. Acad. Sci. USA*, vol. 42, 1956, pp. 920–923.

J. R. Isbell, "On a problem of Robbins," *Ann. Math. Stat.*, vol. 30, 1959, pp. 606–610.

Finally, let us cite some work in the field of economic control processes,

H. Scarf, "Bayes solution of the statistical inventory problem," *Ann. Math. Stat.*, vol. 30, 1959, pp. 490–508.

A. Dvoretzky, J. Kiefer, and J. Wolfowitz, "The inventory problem," *Econometrica*, vol. 20, 1952, pp. 450–466.

CHAPTER XVII

SOME ASPECTS OF
COMMUNICATION THEORY

Wisdom is the principal thing; therefore get
wisdom, and with all thy getting, get understanding.

"*Proverbs*, IV., 7.
Old Testament"

17.1 Introduction

A number of quite interesting problems have arisen in the study of various
methods used for conveying information from one source to another. Not
only are there the many questions involved in transmitting and receiving
with minimal error, or, as in the case of computing machines, with no error,
but there are also the formidable questions of data storage and data
reduction.

In recent years, these matters have assumed particular significance in
the field of biology where the basic reproductive process involves the transmission of instructions whereby one cell creates images of itself. Not only
is this mechanism of prime import in the theory of heredity in general,
but it is now believed that the explanation of the scourge of cancer may lie
in this direction. It remains to determine what mistake in instructions
suddenly causes normal cells to go wildly out of control.

Problems of this type are related to coding theory, an area of great
difficulty in which only a few scattered results have been obtained to date,
and even these at the cost of some effort. Consequently, we shall restrict
our attention to the problem of evaluating a communication system.

We do this by imbedding the communication system within the framework of a multistage decision process, which can then be treated by means
of dynamic programming techniques. This enables us to obtain the "channel
capacity" of information theory, as a special case of more general results.

Both the stochastic version and adaptive version of the control processes
we encounter will be discussed, since this gives us another opportunity to
indicate the application of the formalism of the preceding chapter.

219

17.2 A Model of a Communication Process

Let us begin our discussion by considering a simple mathematical model of one aspect of the communication area. More complex mathematical models can be fashioned along similar lines to take account of more complex physical situations.

Consider a source, S, that produces at discrete times a sequence of signals, contaminated by certain undesirable influences which we call noise. These extraneous signals are usually taken to be stochastic in nature, mainly because not enough is known about them to treat them in a deterministic fashion. The combined signal is fed into the ubiquitous black box which we call a "communication channel," and this, in turn, emits a signal. This is the output signal that we observe.

On the basis of the observation of this signal, we wish to make various deductions concerning the properties of the original signal. Observe how close in essence this problem is to the abstract question treated in the chapter devoted to sequential machines.

Schematically, the process may be represented in the following way:

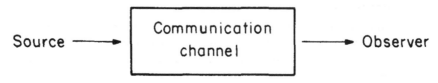

Source ⟶ Communication channel ⟶ Observer

Figure 21

We shall suppose here that there is no feedback between the observer and the source, although in many situations this is an essential feature of the communication process.

In mathematical terms, let

$x =$ a state vector representing the pure signal emanating from S,

$r =$ the noise associated with the signal,

$x' = F(x, r)$, the input to the communication channel, compounded of signal plus noise,

$y =$ the vector signal (not necessarily of the dimension of x) transmitted to the observer by the channel.

Finally, we require a description of the way in which the channel alters the signal x'. The transformation should depend upon the source, and possibly upon the stochastic effects. Let us then write

(17.1) $$y = T[F(x, r), x, r].$$

The analytic problem that arises from this is quite interesting. Given a knowledge of y, we wish to determine as much as possible about the nature of x, the distribution of r, the form of $F(x, r)$, and the structure of T.

This is the problem in its most general and realistic form. Naturally, we shall handle only a small part of it.

17.3 Utility a Function of Use

In order to compare two different communication systems, or parts thereof, we wish to introduce a *metric* into the space of transformations of the form appearing in (17.1). It is rather reasonable to suppose that this can be done in a large number of ways, dependent upon the nature of the source, the type of observer, and upon the personal philosophies involved.

To obtain the desired comparison, we must be able to evaluate the performance of a communication system. In some special cases it is possible to assign individual figures of merit to particular components such as transmitters, receivers, and communication channels without reference to other components of the overall system. Generally, in treating significant processes in a realistic fashion, it is necessary to consider the entire system as a unit.

A communication channel must then be evaluated in terms of the nature of the source feeding into it and the observer which it in turn feeds. Omission of either of these fundamental components leads necessarily to theories of limited validity and applicability.

17.4 A Stochastic Allocation Process

Let us assume that the observer possesses a vector resource, z, the state vector, which can be allocated in various ways depending upon the nature of signal y received from the channel. Let w be the vector allocation.

The effect of this allocation w is to change z into a stochastic vector $R(w, z, y)$, whose distribution we shall take for the moment to be known. This process is now repeated N times, where N may be fixed or may itself be stochastic, dependent upon the signal y and the decision w. Let us consider only the simplest case, that of fixed N.

As usual in a multistage decision process, the purpose of the observer in carrying out this process is to maximize the expected value of some function of the states and decisions. Let us suppose here that it is desired to maximize the expected value of a given function of the final state z_N.

If the channel is distortionless, which is to say, error-free, and if there are no stochastic disturbances, r, the transformation in (17.1) will be the identity transformation. Let f_I denote the maximum expected value of the criterion function obtained in this ideal case, and let f_T denote the maximum expected value in the usual case where there are imperfections in the signal and errors in the transmission.

We then agree to measure the worth of the communication system by comparing f_T and f_I. Let $g(f_T, f_I)$ be the function which effects this comparison. In this fashion, we introduce a metric into the space of transformations T, and thus into the set of communication channels.

Particular examples of this metrization will be discussed in what follows.

17.5 More General Processes

The process described above is a particular representation of a class of decision processes. Another member with features of great interest is the following. The observer is required to make a decision concerning the nature of the original signal x. He can observe as many samples of the signal emitted by the communication system as he wishes, subject to constraints imposed by costs of testing and limitations of time.

Using the information gained in this way, he makes an estimate concerning the properties of the input signal. As a result of this, a cost is sustained dependent upon the deviation of this estimate from the actual signal.

The problem is to combine the testing and estimation in such a way as to minimize the expected total cost, determined by the cost of testing and the cost of deviation. The theory of sequential analysis is devoted to one aspect of this general problem.

Both processes, the one described here and the one described in the preceding section, are deterministic versions of processes in which neither the structure of the communication channel, nor the transformations or probability distributions are completely known. We shall treat a particular adaptive control process of this nature in what follows.

17.6 The Efficient Gambler

Consider a gambler who receives information concerning the outcomes of a sequence of independent sporting events over a noisy communication channel. We shall assume that the outcome of an event is the result of play between two evenly matched teams so that there is no a priori reason for favoring one outcome over the other. The results of a match can, however, be garbled in transmission due to noise. Let p be the probability of correct transmission and $q = 1 - p$, the probability of an incorrect transmission.

Assume that the gambler starts with an initial amount x and uses the advance information to bet on the outcome of each event. If there were no noise, p would equal 1, and the gambler would automatically win each bet. If he places his bets so as to maximize his expected capital at the end of N stages of play, it is clear that his optimal policy is to wager his entire fortune on each play if $p > \frac{1}{2}$, and to wager nothing if $p < \frac{1}{2}$.

This is, however, a risky policy, since even if $p > \frac{1}{2}$, as long as it is less than 1 there is a high probability that the gambler will be ruined pursuing this betting procedure over a multistage process with many stages. Suppose that he wishes to pursue a more conservative policy that will prevent his ever being ruined. Consider then the situation in which he wishes to maximize the expected value of the logarithm of the total amount he possesses at the end of N stages. It is clear that this criterion function rules out any policy which offers a chance of reducing his fortune to zero.

For the one-stage process, we face the problem of maximizing the function

$$(17.2) \qquad E_1(y) = p \log (x + y) + q \log (x - y)$$

over all y in $[0, x]$. Here $q = 1 - p$. It is easy to see that if $p > q$, the maximizing value of y is given by

$$(17.3) \qquad y = (p - q)x,$$

and that this yields as the expected value

$$(17.4) \qquad E_1 = \log x + \log 2 + p \log p + q \log q.$$

If $p \leq q$, the maximum is at $y = 0$.

It is not difficult to show that in an N-stage process where we are constrained by the condition that the same fraction of the capital is wagered at each stage, the policy stated above is optimal. We wish to show that this policy, which is to wager a fixed fraction $p - q$ of the available capital, is optimal within the class of *all* betting policies.

17.7 Dynamic Programming Approach

Following the usual procedure, introduce the sequence of functions

(17.5) $f_N(x) =$ the expected value of the logarithm of the final capital, obtained from an N-stage process starting with the initial amount x and using an optimal policy.

Arguing as before, we obtain the recurrence relation

$$(17.6) \qquad f_N(x) = \underset{0 \leq y \leq x}{\text{Max}} [pf_{N-1}(x + y) + qf_{N-1}(x - y)],$$

$N \geq 2$, with

$$(17.7) \qquad f_1(x) = \log x + K,$$

where

$$(17.8) \qquad K = \begin{cases} \log 2 + p \log p + q \log q, & p > q \\ 0, & p \leq q. \end{cases}$$

Let us show inductively that

$$(17.9) \qquad f_N(x) = \log x + NK,$$

for $N > 1$. Furthermore, the optimal policy is independent of N and has the form mentioned above, namely

$$(17.10) \qquad y = (p - q)x, \quad p \geq q,$$
$$= 0, \quad p \leq q.$$

Assume that the result holds for N. Then

$$(17.11) \quad f_{N+1}(x) = \underset{0 \leq y \leq x}{\text{Max}} \left[p[\log (x + y) + NK] + q[\log (x - y) + NK]] \right]$$

$$= NK + \underset{0 \leq y \leq x}{\text{Max}} \left[p \log (x + y) + q \log (x - y) \right]$$

$$= NK + \log x + K = \log x + (N + 1)K.$$

17.8 Utility of a Communication Channel

If the channel transmits without error, $p = 1$, and $K = \log 2$, the maximum possible value. Thus, the difference between the maximum expected value for a perfect channel and that for a channel in which $p < 1$ is $N[p \log p + q \log q]$.

To obtain a "steady-state" value, we divide by N and obtain a rate, the quantity $p \log p + q \log q$, the magic number which occupies a prominent role in information theory. This quantity is called the "channel capacity" and is often used as a measure of a communication system.

One advantage of the foregoing analysis is that it shows very clearly how special a criterion this function is, and how closely it is bound to the special function $\log x$. Change this function and one changes the measure obtained above.

17.9 Time-dependent Case

Let us now consider the case in which the channel properties change with time. Suppose that at the k-th stage the probability of correct transmission is p_k, and of incorrect transmission $q_k = 1 - p_k$. As before, we wish to maximize the expected value of the logarithm of the capital remaining after N stages.

Let

$(17.12) \quad f_k(x) =$ the expected value of the logarithm of the final capital obtained from the remaining k stages of the original N-stage process, starting with initial capital x and using an optimal policy.

Then

$$(17.13) \quad f_1(x) = \underset{0 \leq y \leq x}{\text{Max}} \left[p_N \log (x + y) + q_N \log (x - y) \right],$$

$$f_k(x) = \underset{0 \leq y \leq x}{\text{Max}} \left[p_{N-k+1} f_{k-1}(x + y) + q_{N-k+1} f_{k-1}(x - y) \right],$$

for $N \geq k \geq 2$.

As before, it follows inductively that

$$(17.14) \quad f_k(x) = \log x + k \log 2 + \sum_{r=N-k+1}^{N} [p_r \log p_r + q_r \log q_r],$$

provided that $p_k > \frac{1}{2}$ for all k. Whenever this condition fails, the expression $p_k \log p_k + q_k \log q_k$ must be replaced by $-\log 2$.

17.10 Correlation

Let us now consider the case where the stochastic effects within the channel are not independent. A simple example of this realistic phenomenon is that where the probability of transmission at a particular stage depends upon whether or not the preceding signal was transmitted correctly. Although a large variety of questions of this type may be formulated, we feel that the following discussion will be sufficient to illustrate a uniform method that may be employed to treat questions of this nature.

Let

p_k = the probability of correct transmission of the k-th signal, provided that the $(k-1)$-th signal was transmitted correctly.

r_k = the probability of correct transmission of the k-th signal, provided that the $(k-1)$-th signal was transmitted incorrectly.

As before, introduce the sequence of functions

$f_k(x)$ = the expected value of the logarithm of the final capital obtained from the remaining k stages of the original N-stage process, starting with an initial quantity x and the information that the $(k-1)$-th signal was transmitted correctly, and using an optimal policy.

$g_k(x)$ = the corresponding function in the case where the $(k-1)$-th signal was transmitted incorrectly.

As above, we derive the recurrence relations

$$(17.15) \quad f_k(x) = \underset{0 \le y \le x}{\text{Max}} \left[p_{N-k+1} f_{k-1}(x+y) + (1 - p_{N-k+1}) g_{k-1}(x-y) \right],$$
$$g_k(x) = \underset{0 \le y \le x}{\text{Max}} \left[r_{N-k+1} f_{k-1}(x+y) + (1 - r_{N-k+1}) g_{k-1}(x-y) \right].$$

It follows, inductively again, that

$$(17.16) \qquad\qquad f_k(x) = \log x + a_k,$$
$$g_k(x) = \log x + b_k,$$

where the recurrence relations for the a_k and b_k are readily established.

17.11 *M*-signal Channels

Let us now consider the more interesting situation in which the channel is called upon to transmit any of M different symbols, say the integers $1, 2, \ldots, M$. Upon receiving a particular symbol j, the gambler is free to make bets on what he believes the original signal to have been.

We suppose that the gambler possesses the following information:

$p_{ij} =$ the conditional probability that the j-signal was emitted at the source if the i-signal is received by the observer.

$q_i =$ the probability that the observer at any stage will observe an i-signal.

$r_j =$ the return from a winning bet of one unit on signal j.

When the gambler receives the j-signal, he bets an amount z_i that it was actually the i-signal, subject to the restrictions

$$(17.17) \qquad \sum_{i=1}^{M} z_i \le x, \quad z_i \ge 0.$$

As before, we assume that the gambler proceeds so as to maximize the expected value of the logarithm of the capital he possesses at the end of N stages.

Defining the sequence $\{f_N(x)\}$ as above, we obtain the relations

$$(17.18) \quad f_N(x) = \sum_{i=1}^{M} q_i \left\{ \operatorname*{Max}_{\substack{\Sigma z_j \le x \\ z_j \ge 0}} \left[\sum_{j=1}^{M} p_{ij} f_{N-1}\left(r_j z_j + x - \sum_{s=1}^{M} z_s\right) \right] \right\}, \quad N \ge 2,$$

and

$$(17.19) \quad f_1(x) = \sum_{i=1}^{M} q_i \left\{ \operatorname*{Max}_{\substack{\Sigma z_j \le x \\ z_j \ge 0}} \left[\sum_{j=1}^{M} p_{ij} \log \left(r_j z_j + x - \sum_{s=1}^{M} z_s\right) \right] \right\}.$$

Once again, it is not difficult to prove inductively that

$$(17.20) \qquad\qquad f_N(x) = \log x + NK,$$

where

$$(17.21) \qquad K = \sum_{i=1}^{M} q_i \left\{ \operatorname*{Max}_{\substack{\Sigma z_j \le 1 \\ z_j \ge 0}} \left[\sum_{j=1}^{M} p_{ij} \log \left(r_j z_j + 1 - \sum_{s=1}^{M} z_s\right) \right] \right\}.$$

From the expression for K, we note that the optimal policy depends only upon the quantities p_{ij} and r_j, although the return itself depends upon the q_i. An interesting special case is that where the gambler is required to allocate all of his funds at each stage, $\sum_{i=1}^{M} z_i = x$. In this case, it is easily seen that the optimal policy depends only upon the quantities p_{ij}. In other words, the properties of the source, and the properties of the associated decision process, play no role in determining the utility of the channel.

This is, of course, a rather unrealistic result, and one that must be used with care.

In order to obtain the relation between the source characteristics and those of the channel, let

$p_i =$ the probability of sending an i-signal from S,

$t_{ij} =$ the conditional probability that if an i-signal is sent, then a j-signal is received by the observer.

Then

(17.22)
$$p_i = \sum_{j=1}^{M} q_j p_{ji},$$

$$q_i = \sum_{j=1}^{M} p_j t_{ji},$$

which means that the matrices (t_{ij}) and (p_{ij}) are inverses.

17.12 Continuum of Signals

Consider the case where there are a continuum of different signals emitted by S. Let

$dG(u, v) =$ the conditional probability that a signal with label between v and $v + dv$ is sent if the u-signal is received,

$$-\infty < u, v < \infty,$$

and let

$dH(u) =$ the probability that a signal with label between u and $u + du$ is received at any stage.

Restricting ourselves to the case of even odds, for the sake of simplicity, we derive the recurrence relations

(17.23)
$$f_N(x) = \int_{-\infty}^{\infty} \left[\underset{z(v)}{\text{Max}} \int_{-\infty}^{\infty} f_{N-1}[2z(v)] \, dG(u, v) \right] dH(u), \quad N \geq 2,$$

$$f_1(x) = \int_{-\infty}^{\infty} \left[\underset{z(v)}{\text{Max}} \int_{-\infty}^{\infty} \log \left[2z(v) \right] \, dG(u, v) \right] dH(u).$$

In both cases, maximization is over all functions $z(v)$ satisfying the conditions

(17.24) (a) $z(v) \geq 0,$

(b) $\displaystyle\int_{-\infty}^{\infty} z(v) \, dv = x.$

Once again, it is easily seen that

(17.25)
$$f_N(x) = \log 2x + KN,$$

where

(17.26)
$$K = \int_{-\infty}^{\infty} \underset{w(v)}{\text{Max}} \left[\int_{-\infty}^{\infty} \log w(v) \, dG(u, v) \right] dH(u),$$

with $w(v)$ subject to the conditions

(17.27) (a) $w(v) \geq 0,$

(b) $\displaystyle\int_{-\infty}^{\infty} w(v) \, dv = 1.$

227

17.13 Random Duration

Let us now consider the case where the duration of the process is a stochastic quantity dependent upon the sequence of decisions and events. To simplify the notation, we consider only the original simple model.

Let

$b(y)$ = the probability that the process terminates at the next stage if the observer bets y.

Define

$f(x)$ = the expected value of $\phi(z)$, where z is the amount of money at the termination of the process.

Then,

$$(17.28) \quad f(x) = \underset{0 \leq y \leq x}{\text{Max}} [b(y)[p\phi(x + y) + q\phi(x - y)]$$
$$+ [1 - b(y)][pf(x + y) + qf(x - y)]].$$

As in a number of processes treated in the preceding pages, some method of successive approximations must be used to determine $f(x)$.

17.14 Adaptive Processes

Let us now consider a situation in which the stochastic properties of the channel are not completely known.

Assume that the source emits two types of signals, say 0's and 1's. Passing through the communication channel, there is a probability p that there will be correct transmission, i.e. that a 0 is transformed into a 0 and a 1 into a 1. With probability $q = 1 - p$, a 0 is transformed into a 1 or a 1 into a 0.

Consider the case in which it is known that p has a fixed value, but where this value is unknown. As the process continues, however, we will be able to make better and better estimates of p. We begin, as before, by postulating both an a priori distribution function for p, $dG(p)$, and a means of transforming $dG(p)$ on the basis of experience.

If we observe m successful wagers and n unsuccessful wagers, we change $dG(p)$ into

$$(17.29) \qquad dG_{m,n}(p) = \frac{p^m(1 - p)^n \, dG(p)}{\displaystyle\int_0^1 p^m(1 - p)^n \, dG(p)}.$$

Let

$f_N(x; m, n)$ = the expected value of $\phi(z_N)$, where z_N is the capital at the end of N stages, starting with a quantity x and the information that there have been m successes and n failures to date, and using an optimal policy.

Introduce the notation

$$(17.30) \qquad p_{mn} = \frac{\int_0^1 p^{m+1}(1-p)^n \, dG}{\int_0^1 p^m(1-p)^n \, dG},$$

the expected probability of correct transmission. Then we have the following recurrence relation:

$$(17.31) \quad f_N(x; m, n) = \operatorname*{Max}_{0 \leq y \leq x} [p_{mn} f_{N-1}(x + y; m + 1, n)$$

$$+ (1 - p_{mn}) f_{N-1}(x - y; m, n + 1)],$$

$N = 2, 3, \ldots,$ with

$$(17.32) \quad f_1(x; m, n) = \operatorname*{Max}_{0 \leq y \leq x} [p_{mn} \phi(x + y) + (1 - p_{mn}) \phi(x - y)].$$

If $\phi(z) = \log z$, it is easy to show that

$$(17.33) \qquad f_N(x; m, n) = \log x + c_N(m, n),$$

where

$$(17.34) \quad c_N(m, n) = p_{mn} c_{N-1}(m + 1, n) + (1 - p_{mn}) c_{N-1}(m, n + 1)$$

$$+ \operatorname*{Max}_{0 \leq y \leq 1} [p_{mn} \log (1 + y) + (1 - p_{mn}) \log (1 - y)].$$

Although it is not easy to evaluate $c_N(m, n)$ explicitly, the structure of the optimal policy is still remarkably simple:

$$(17.35) \qquad y = (2p_{mn} - 1)x \quad \text{if } p_{mn} > \tfrac{1}{2},$$

$$= 0, \quad \text{otherwise.}$$

For the case where we choose as an initial distribution

$$(17.36) \qquad dG(p) = \frac{p^{a-1}(1 - p)^{b-1} \, dp}{B(a, b)},$$

where $B(a, b)$ is the Beta function, a choice which allows great flexibility in the form of dG, it is easy to see that

$$(17.37) \quad p_{mn} = \frac{B(m + a + 1, n + b)}{B(m + a, n + b)} = \frac{(m + a)}{(m + a) + (n + b)}.$$

For the logarithmic criterion, this yields as the optimal policy after m wins and n losses the procedure of betting the fraction

$$(17.38) \qquad \frac{(m + a) - (n + b)}{(m + a) + (n + b)}$$

of one's capital, provided that $p_{mn} > \tfrac{1}{2}$, and nothing otherwise.

Bibliography and Comments

17.1 A fundamental paper in the field is

C. Shannon, "A mathematical theory of communication," *Bell System Technical J.*, vol. 27, 1948, pp. 379–423 and pp. 623–656.

See also

N. Wiener, "What is information theory," *IRE Trans. on Information Theory*, vol. IT-2, 1956, p. 48.

17.2 Here, and in the remainder of the chapter, we follow the presentation given in

R. Bellman and R. E. Kalaba, "Dynamic programming and statistical communication theory," *Proc. Nat. Acad. Sci. USA*, vol. 43, 1957, pp. 749–751.

R. Bellman and R. E. Kalaba, "On the role of dynamic programming in statistical communication theory," *IRE Trans. on Information Theory*, vol. IT-3, 1957, pp. 197–203.

R. Bellman and R. E. Kalaba, "On communication processes involving learning and random duration," *IRE National Convention Record*, part IV, 1958, pp. 16–20.

17.5 For a discussion of problems of this general nature by other means, see

J. Busgang and D. Middleton, "Optimum sequential detection of signals in noise," *IRE Trans. on Information Theory*, vol. IT-1, 1955, pp. 5–18.

A. Dvoretzky, J. Kiefer, and J. Wolfowitz, "The inventory problem—II: case of unknown distribution of demand," *Econometrica*, vol. 20, 1952, pp. 450–466.

A. Wald, *Sequential Analysis*, John Wiley and Sons, New York, 1947.

17.6 The novel idea of imbedding the problem of determining the utility of a communication channel within a stochastic allocation process is due to J. Kelly. See

J. Kelly, "A new interpretation of information rate," *Bell System Technical J.*, vol. 35, 1956, pp. 917–926.

In this paper, he considers only the special case of wagering a fixed fraction of the available capital at each stage. The general solution given here is contained in the papers of Bellman and Kalaba cited above.

An extensive discussion of this idea, which is quite natural from the standpoint of mathematical economics, is contained in

J. Marshak, *Remarks on the Economics of Information*, Cowles Foundation Discussion Paper No. 70, April, 1959.

See also

C. B. McGuire, *Comparisons of Information Structure*, Cowles Foundation Discussion Paper No. 71, April, 1959,

where references to related results of Blackwell and Bohnenblust-Shapley-Sherman will be found.

17.14 One can also ask what criterion functions, apart from the logarithm, allow invariant policies to be optimal. This question leads to an interesting functional equation which is discussed in the second of the papers by Bellman and Kalaba cited above.

For further results, see

M. Marcus, "The utility of a communication channel and applications to suboptimal information handling procedure," *IRE Trans. on Information Theory*, vol. IT-4, 1958, pp. 147–151.

For many further references, see

F. L. H. M. Stumpers, "A bibliography of information theory," *IRE Trans.*, PGIT-2; Nov. 1953, First Supplement in vol. IT-1; Sept. 1955; Second Supplement in vol. IT-3; June, 1957.
(I wish to thank P. Metzelaar for these references.)

and

R. Kalaba, "On some communication network problems," *Ninth Symposium on Appl. Math. Proc.*, Amer. Math. Soc., 1960.

For a new approach to prediction theory based upon dynamic programming, see

R. E. Kalman, "On a new approach to filtering and prediction," *J. Basic Engineering*, March 1960, Part D, ASME Trans.

T. Odanaka, "Prediction Theory and Dynamic Programming," The International Statistical Institute, 32 Session, 30/5–9/6, 1960, Tokyo.

CHAPTER XVIII

SUCCESSIVE APPROXIMATIONS

Ουδέ τι οἶδε νοῆσαι ἄμα προσσω καὶ ὀπισσω
. . . Into the Past and Future.

HOMER, "*Iliad*," I, 343

18.1 Introduction

Looking back over what has preceded, we see that although we now possess some techniques which permit us to formulate a variety of control processes in precise mathematical terms, the dimensionality difficulties we have previously dwelt upon effectively prevent us from obtaining routine computational solutions. Since one of our avowed aims is to accomplish this task, we must develop more powerful numerical methods.

As usual, we obtain these by invoking more sophisticated and ingenious analytic techniques. These will be consequences of a combination of the functional equation method used throughout and that general factotum of analysis to which we have already referred a number of times—the *method of successive approximations*, the great legacy to analysis by E. Picard.

Classically, the method of successive approximations is based upon the idea of solving a succession of equations, or problems, which converge to the desired equation, or problem. The dynamic programming approach yields still another path by way of approximation in policy space, a very powerful and natural technique which does not exist in classical analysis. In the chapter on quasilinearization, we showed how this idea could be adapted to the study of some of the usual functional equations of analysis.

As we shall see, the basic idea is a simple one—when once pointed out. What does require the attributes of a mathematician, ingenuity, experience, and divination, is the expeditious application of this idea. Generally, one wishes to apply all of the special devices at one's command. One fact is certainly clear, the efficient treatment of linear control processes with quadratic criteria will be essential.

As we shall see, even in connection with problems of this nature where explicit analytic solutions exist, the functional equation approach yields

232

in many important cases another explicit analytic solution which yields a superior computational algorithm.

In what follows, we shall discuss a number of ways in which the technique of successive approximations can be applied to various types of variational problems. Since this is a new field of research, we shall content ourselves with the formal aspects and deterministic processes alone, leaving aside considerations of rigor and applications to adaptive and stochastic processes.

In many cases, however, it is easy to establish monotonicity of approximation, if not convergence. This brings up the interesting point that in posing variational problems, there are two distinct types of questions to consider. One is that of determining the optimizing policy; the other is that of obtaining a control policy which yields a better result than that currently in use. Often, indeed, practical considerations require a gradual change from one policy to the other. Several of the methods we present are ideally suited to this purpose.

18.2 The Classical Method of Successive Approximations

Let us begin discussing the application of the method of successive approximations to the solution of functional equations. Consider, for example, the scalar differential equation

$$(18.1) \qquad \frac{du}{dt} = g(u), \quad u(0) = c.$$

Let $u_0(t)$ be an initial guess, and let $u_1(t)$ be determined as the solution of the differential equation

$$(18.2) \qquad \frac{du_1}{dt} = g(u_0), \quad u_1(0) = c,$$

and, generally, let us determine a sequence of functions $\{u_n(t)\}$ by means of the recurrence relation

$$(18.3) \qquad \frac{du_n}{dt} = g(u_{n-1}), \quad u_n(0) = c.$$

The differential equation in (18.3) cum boundary condition can be replaced by the integral equation

$$(18.4) \qquad u_n = c + \int_0^t g(u_{n-1}) \, ds.$$

Under reasonable assumptions concerning the function $g(u)$ it can be shown that the sequence generated in this way converges uniformly to the solution of (18.1) in some interval $[0, t_0]$.

Since the original differential equation can be written in the form of an integral equation

$$(18.5) \qquad u = c + \int_0^t g(u) \, ds,$$

we can describe what we have been doing in the following general terms. Consider the equation

$$(18.6) \qquad u = T(u),$$

where $T(u)$ is a transformation of u, such as, for example, that appearing on the right-hand side in (18.5).

Let u_0 be an initial guess, and determine u_1 by means of the relation $u_1 = T(u_0)$. Continuing in this way, we generate the sequence $\{u_n\}$ by means of the relation

$$(18.7) \qquad u_n = T(u_{n-1}).$$

Provided that $T(u)$ satisfies certain conditions, we can show that u_n converges as $n \to \infty$ to a solution u of (18.6). For example, we may require that

$$(18.8) \qquad \|T(u) - T(v)\| \le k\|u - v\|,$$

where $0 < k < 1$, and $\| \ldots \|$ indicates a norm in a suitable space.

In general, a certain amount of ingenuity is required to convert a given equation into a form in which the foregoing general technique can be applied. For example, if we wish to study the solution to

$$(18.9) \qquad \frac{du}{dt} = -u + u^2, \quad u(0) = c,$$

in the neighborhood of $t = 0$, we can write

$$(18.10) \qquad u = c + \int_0^t (u^2 - u) \, ds.$$

If, however, we wish to study the solution in the neighborhood of $t = \infty$, we write

$$(18.11) \qquad \frac{du}{dt} + u = u^2, \quad u(0) = c,$$

and thus, regarding u^2 as a forcing function, are led to the integral equation

$$(18.12) \qquad u = ce^{-t} + \int_0^t e^{-(t-s)} u^2(s) \, ds.$$

18.3 Application to Dynamic Programming

Consider now an equation of the form

$$(18.13) \qquad f(p) = \operatorname*{Max}_q \, [f[T(p, q)] + h(p, q)],$$

which we have met on several occasions. Write

$$(18.14) \qquad f_0(p) = \operatorname*{Max}_q h(p, q),$$

and, generally,

$$(18.15) \qquad f_{n+1}(p) = \operatorname*{Max}_q \, [h(p, q) + f_n[T(p, q)]].$$

This corresponds to the approximation of an unbounded process by a sequence of processes, each of a fixed number of stages.

We need, however, not use this type of approximation, which may converge quite slowly. We may choose $f_0(p)$ in some other way and then determine the remaining elements of the sequence $\{f_n(p)\}$ by way of (18.15). One advantage of the scheme in (18.14) and (18.15) lies in the fact that

$$f_0(p) \leq f_1(p) \leq \ldots \leq f_n(p) \leq f_{n+1}(p) \leq \ldots \quad \text{if} \quad h(p, q) \geq 0.$$

On the other hand, a method based upon an arbitrary choice of $f_0(p)$ may converge, but not monotonically.

In the next section, we shall describe another method of successive approximation which always yields monotone convergence.

18.4 Approximation in Policy Space

Let us begin with the observation that dynamic programming processes generate two types of functions, return functions and policy functions. As far as the actual decision process is concerned, it is far more natural to think in terms of an approximate *policy* rather than in terms of an approximate *function*. Furthermore, particularly in adaptive control processes, policies may be well-defined in many situations in which the criterion function is not precisely defined.

Let us then think in terms of an approximation in policy space. In place of guessing an initial function $f_0(p)$, let us guess an initial policy, $g_0(p)$. This policy determines a return function $f_0(p)$ by means of the functional equation

$$(18.16) \qquad f_0(p) = f_0[T(p, q_0)] + h(p, q_0),$$

which we solve by means of iteration.

For example, if the functional equation were

$$(18.17) \qquad f(x) = \max_{0 \leq y \leq x} [g(y) + h(x - y) + f[ay + b(x - y)]],$$

$0 < a < b < 1$, $x \geq 0$, we may take as a simple initial policy

$$(18.18) \qquad y_0(x) = 0, \quad x \geq 0.$$

Then, supposing that $g(0) = 0$, we have

$$(18.19) \qquad f_0(x) = h(x) + f_0(bx).$$

Iteration yields

$$(18.20) \qquad f_0(x) = h(x) + h(bx) + h(b^2 x) + \ldots.$$

Returning to (18.16), having determined $f_0(p)$, we now determine $q_1(p)$ as the function yielding the maximum in

$$(18.21) \qquad \max_q [f_0[T(p, q)] + h(p, q)].$$

Using $q_1(p)$, we determine a new return function $f_1(p)$ by means of the relation

(18.22) $$f_1(p) = f_1[T(p, q_1)] + h(p, q_1).$$

We now wish to compare the two functions $f_1(p)$ and $f_0(p)$. We have

(18.23) $$f_0(p) = f_0[T(p, q_0)] + h(p, q_0) \leq \underset{q}{\text{Max}} \, [f_0[T(p, q)] + h(p, q)]$$
$$= f_0[T(p, q_1)] + h(p, q_1).$$

We will be able to assert the inequality $f_0(p) \leq f_1(p)$, provided that it is true that any solution of the inequality

(18.24) $$u(p) \leq u[T(p, q_1)] + h(p, q_1),$$

with $q_1 = q_1(p)$, is dominated by the solution of the corresponding equality, the function $f_1(p)$. This property will generally hold for the equations of dynamic programming.

Continuing in this fashion, we obtain a sequence of functions $\{f_n(p)\}$ which, under the foregoing assumption of the positivity of the operator $u[T(p, q)] - u(p) + h(p, q)$, are monotone increasing in n. Furthermore, since a choice of $q_0(p)$ yields a return function which is at most equal to that derived from the optimal policy, we see that $f_0(p) \leq f(p)$, and similarly, $f_n(p) \leq f(p)$. Hence,

(18.25) $$f_0(p) \leq f_1(p) \leq \ldots \leq f_n(p) \leq \ldots \leq f(p).$$

Consequently, we see that approximation in policy space yields monotone approximation which is actually monotone convergence.

18.5 Quasilinearization

Comparing the foregoing abstract formulation of approximation in policy space with the techniques presented in the chapter on quasilinearization, we see that these latter techniques are particular applications of the general concept. Not only is it true that the equations of dynamic programming possess the required positivity property, but of much wider import is the fact that many of the basic equations of mathematical physics possess this same property.

18.6 Application of the Preceding Ideas

Consider, as an illustration of several of the foregoing ideas, the problem of transforming a system from an arbitrary initial state p to a prescribed state p_0 in minimum time. At each stage we have a choice of one of a set of transformations $T(p, q)$, each of which consumes a unit of time. In the usual fashion, we obtain an equation

(18.26) $$f(p) = 1 + \underset{q}{\text{Min}} \, f[T(p, q)].$$

Generally, in order to solve this equation, we must employ the method of successive approximations in some way.

236

The first way is to guess a function $f_0(p)$ and use this as the first member of a sequence $\{f_n(p)\}$ determined by the recurrence relation

(18.27) $$f_{n+1}(p) = 1 + \operatorname*{Min}_{q} [f_n[T(p, q)]].$$

In place of doing this, we regard the original problem as a limiting form of a sequence of problems and obtain the solution to the original problem as the limit of the solutions of this sequence of problems.

As an example of this, consider the problem of minimizing the distance from p_0 in time N. Then, if $\|p - p_0\|$ represents the distance, we have

(18.28) $$f_0(p) = \|p - p_0\|,$$
$$f_N(p) = \operatorname*{Min}_{q} [f_{N-1}[T(p, q)]].$$

The smallest N for which $f_N(p) = 0$ is $f(p)$. We have mentioned this technique above in several places.

If the original process is unbounded, we can use finite processes for approximation purposes. Conversely, if the original process is bounded, we can use unbounded processes for approximation purposes.

To approximate in policy space, we guess an initial policy $q_0 = q_0(p)$ and determine $f_0(p)$ by means of the relation

(18.29) $$f_0(p) = 1 + f_0[T(p, q_0)].$$

We can now either continue by means of the sequence

(18.30) $$f_{n+1}(p) = 1 + \operatorname*{Min}_{q} f_n[T(p, q)],$$

or by means of the relation

(18.31) $$f_1(p) = 1 + f_1[T(p, q_1)],$$

where $q_1 = q_1(p)$ is, as in 18.4, a function which maximizes $f_0[T(p, q)]$.

The advantage of approximation in policy space over the usual approximation methods lies in the fact that we may begin with an initial approximation which is far closer to the solution, an approximation arrived at on the basis of intuition and experience.

18.7 Successive Approximations in the Calculus of Variations

Since problems in the calculus of variations can be interpreted as dynamic programming processes, we know that we can obtain a variety of approximate techniques by specializing the foregoing general techniques. Let us now, however, discuss some analytic devices particularly adapted to the study of variational problems involving differential equations and functionals of integral form.

To begin with, let us note that we can use approximation techniques either to reduce the problem to a form in which it permits a reasonably efficient analytic solution or to transform it in such a way as to allow an efficient computational algorithm. We have emphasized throughout this volume that a characteristic of more realistic versions of classical processes and of some of the new processes arising in various areas of scientific study is that a great deal of effort and ingenuity is required both to formulate and obtain numerical solutions.

Let us give examples of both types of approximations, using the problem of minimizing

$$(18.32) \qquad J(x) = \int_0^T g(x_1, x_2, \ldots, x_N; x_1', x_2', \ldots, x_N') \, dt$$

in both cases. Let x^0 be an initial approximation, and let us expand $g(x_1, x_2, \ldots, x_N; x_1', x_2', \ldots, x_N')$ around this trial solution, retaining only the terms up to and including the quadratic. The new variational problem is that of minimizing

(18.33)

$$J_1(x) = J[x^0] + \int_0^T [x - x^0, b_0(t)] \, dt + \int_0^T [x - x^0, A_0(t)(x - x^0)] \, dt,$$

where $b_0(t)$ and $A_0(t)$ are respectively a known vector and matrix determined by x^0.

This quadratic variational problem can be resolved explicitly to yield a vector x^1. Repeating this procedure, we obtain a sequence of vectors $\{x^n\}$ which we hope converges to the solution. Problems of this interesting and important nature have not as yet been investigated.

Let us now indicate how we can approach the general problem by a sequence of problems which can be treated by functional equation techniques in terms of sequences of functions of one variable.

Let $x^0 = [x_1^0, x_2^0, \ldots, x_N^0]$ be an initial guess and consider the new problem of minimizing

$$(18.34) \qquad J_1(x) = \int_0^T g[x_1, x_2^0, \ldots, x_N^0; x_1', (x_2^0)', \ldots, (x_N^0)'] \, dt.$$

This requires only sequences of functions of one variable, since only x_1 is free to vary. Consequently, we can consider this to be a routine task. Let x_1^1 be the function obtained in this way and write

$$(18.35) \qquad x^1 = [x_1^1, x_2^0, \ldots, x_N^0].$$

Clearly

$$(18.36) \qquad J(x^0) \geq J(x^1).$$

Now require that $x_1, x_3, \ldots, x_N$ assume the functional values $x_1^1, x_3^0, \ldots, x_N^0$, respectively, and consider the new problem of minimizing

$$(18.37) \qquad J_2(x) = \int_0^T g[x_1^1, x_2, x_3^0, \ldots, x_N^0; (x_1)', x_2, \ldots, x_N^0] \, dt,$$

where only x_2 is allowed to vary. Let x_2^2 be the function determined in this way. Since x_2^0 was a feasible choice of x_2, we see that

$$(18.38) \qquad J(x^2) \leq J(x^1) \leq J(x^0),$$

where

$$(18.39) \qquad x^2 = [x_1^1, x_2^2, x_3^0, \ldots, x_N^0].$$

Continuing in this way, we see that we obtain monotone approximation. Although the problem of convergence is, as might be expected, quite complex, we have achieved our goal of obtaining an approximation method which improves upon any trial solution.*

18.8 Preliminaries on Differential Equations

We shall require for our further results some well-known results concerning the solution of linear systems of the form

$$(18.40) \qquad \frac{dx}{dt} = Ax + y, \quad x(0) = c,$$

where A is a constant matrix. Let $X(t)$ denote the solution of the matrix equation

$$(18.41) \qquad \frac{dX}{dt} = AX, \quad X(0) = I,$$

sometimes written as the matrix exponential e^{At}. Then the solution of (18.40) has the form

$$(18.42) \qquad x = e^{At}c + \int_0^t e^{A(t-s)}y(s)\, ds,$$

in striking analogy to the scalar case—and derived in exactly the same way. Hence,

$$(18.43) \qquad x(T) = e^{AT}c + \int_0^T e^{A(T-s)}y(s)\, ds$$

$$= b + \int_0^T K(s)y(s)\, ds,$$

where b and $K(s)$ are fixed if T is fixed. Thus, each component of $x(T)$ has the form

$$(18.44) \quad x_i(T) = b_i + \int_0^T \left[\sum_{j=1}^N k_{ij}(s)y_j(s) \right] ds, \quad i = 1, 2, \ldots, N.$$

* Recall the duet sung by Frank Butler and Annie Oakley in "Annie, Get Your Gun."

18.9 A Terminal Control Process

Let us now consider the problem of determining the maximum of a prescribed function of the terminal state

$$(18.45) \qquad g[x_1(T), x_2(T), \ldots, x_k(T)]$$

subject to the constraint

$$(18.46) \qquad \int_0^T h(y_1, y_2, \ldots, y_N) \, dt \le k_1,$$

where x and y are related as in (18.40). Using a Lagrange multiplier, we consider the problem of maximizing the functional

$$(18.47) \quad J(y) = g[x_1(T), x_2(T), \ldots, x_k(T)] + \lambda \int_0^T h(y_1, y_2, \ldots, y_N) \, dt.$$

If N is large, the direct functional equation approach is not practical. Let us now show that if k is small, $k = 1, 2, 3$, we can still use the functional equation approach, applied in a different fashion.

Using the representation of (18.44), we can eliminate all dependence upon x and write $J(y)$ directly in terms of y,

$$(18.48) \quad J(y) = g\left[b_1 + \int_0^T \left\{ \sum_{j=1}^N k_{1j}(s)y_j(s) \right\} ds, \ldots, \right.$$

$$\left. b_k + \int_0^T \left\{ \sum_{j=1}^N k_{rj}(s)y_j(s) \right\} ds \right] + \lambda \int_0^T h(y_1, y_2, \ldots, y_N) \, dt.$$

To treat this problem by means of dynamic programming techniques, we consider the problem of maximizing the expression

$$(18.49) \quad J(y, a) = g\left[b_1 + \int_a^T \left\{ \sum_{j=1}^N k_{1j}(s)y_j(s) \right\} ds, \ldots, \right.$$

$$\left. b_k + \int_a^T \left\{ \sum_{j=1}^N k_{rj}(s)y_j(s) \right\} ds \right] + \lambda \int_a^T h(y_1, y_2, \ldots, y_N) \, dt.$$

We regard the lower limit a as variable in the interval $[0, T]$, and the b_i as independent variables constituting new state variables. Then,

$$(18.50) \qquad \underset{y}{\text{Max}} \, J(y, a) = f(b_1, b_2, \ldots, b_k, a).$$

To obtain a functional equation for this new function f, we proceed as in the previous chapters. The result is

$$(18.51) \quad f(b_1, \ldots, b_k, a) = \underset{[v_1, v_2, \ldots, v_N]}{\text{Max}} \left[\lambda \Delta h(v_1, v_2, \ldots, v_N) \right.$$

$$\left. + f\left(b_1 + \Delta \sum_{j=1}^N k_{1j}v_j, \ldots, b_k + \Delta \sum_{j=1}^N k_{rj}v_j, \, a + \Delta \right) \right] + o(\Delta).$$

This can either be used as the basis of a finite difference scheme, or to derive useful analytic relations, as will be indicated below.

18.10 Differential-difference Equations and Retarded Control

In many physical processes where time-lags occur, the equations governing the system are not differential equations, but functional equations of more complex type. As a simple example of an equation of this nature, consider the *differential-difference equation*

(18.52) $$u'(t) = au(t) + bu(t-1) + v(t), \quad t > 1.$$

To determine the solution, we now need an interval initial condition,

(18.53) $$u(t) = g(t), \quad 0 \le t \le 1.$$

To attempt to solve feedback control problems, such as that of minimizing

(18.54) $$|u(T)| + \lambda \int_1^T v^2(s)\, ds,$$

by a direct application of functional equation techniques is clearly fruitless, since it would involve functions of functions. The minimum value of the functional in (18.54) is itself a functional of the initial function $g(t)$.

If functions of many variables cause us computational headaches, any direct computational approach to functionals is unthinkable. Fortunately, terminal control problems can be handled for these more complex processes in precisely the manner outlined in the preceding section. This is a consequence of the fact that the Laplace transform method readily yields a solution of (18.52) in the form

(18.55) $$u = u_0 + \int_1^t k(t-s)v(s)\, ds,$$

where u_0 is the solution of the homogeneous equation.

18.11 Quadratic Criteria

We have outlined in the previous sections a new functional equation approach to the problem of minimizing the functional

(18.56) $$g[x_1(T), \ldots, x_r(T)] + \lambda \int_0^T h(y_1, y_2, \ldots, y_N)\, dt,$$

over all y when x and y are connected by a linear differential equation

(18.57) $$\frac{dx}{dt} = Ax + y, \quad x(0) = c.$$

The new dimension of the process is r. It follows that if $r = 1$ or 2, we have a routine solution. If r is larger we must employ special techniques, one of which we will point out below.

Let us now consider the special case in which g and h are quadratic functions of their arguments. In this case, the variational problem, following classical lines, leads to the problem of solving a system of linear differential equations of order $2N$, where N is the dimension of x, with a two-point boundary condition, at $t = 0$ and T. We shall show that if r is substantially smaller than N, then the functional equation technique yields a nonlinear algorithm which is computationally superior to the linear algorithm described immediately above.

For example, if $N = 100$ and $r = 10$, the nonlinear method will be preferable. Essentially, this will be due to the fact that the functional equation technique is an initial value method.

It is clear from the linearity of the Euler equation that the solution, given by the functions $x(t)$ and $y(t)$, depends in a linear function upon the initial variables. Hence, the function $f(b_1, b_2, \ldots, b_k; a)$ used in 18.9 is a quadratic form in the variables b_i, with coefficients that depend upon a,

$$(18.58) \qquad f(b_1, b_2, \ldots, b_k; a) = \sum_{i,j=1}^{r} q_{ij}(a)b_i b_j.$$

The limiting form of (18.51) as $\Delta \to 0$ is the partial differential equation

$$(18.59) \qquad -\frac{\partial f}{\partial a} = \underset{v}{\mathrm{Min}} \left[\lambda h(v_1, v_2, \ldots, v_N) + \sum_{i=1}^{r} \left[\sum_{j=1}^{N} k_{ij} v_j \right] \frac{\partial f}{\partial b_i} \right].$$

We now assert that the right-hand side reduces to a quadratic function of the partial derivatives $\partial f/\partial b_i$ if $h(v_1, v_2, \ldots, v_N)$ is a positive definite quadratic form in the v_i.

Let

$$(18.60) \qquad h(v_1, v_2, \ldots, v_N) = (v, Cv).$$

Then we wish to compute an expression of the form

$$(18.61) \qquad \underset{v}{\mathrm{Min}} \left[\lambda(v, Cv) + 2(v, z) \right],$$

where the components of z are linear functions of the partial derivatives, i.e.

$$(18.62) \qquad z = Q \begin{pmatrix} \dfrac{\partial f}{\partial b_1} \\ \cdot \\ \cdot \\ \cdot \\ \dfrac{\partial f}{\partial b_r} \end{pmatrix},$$

where the matrix Q is determined by the k_{ij} appearing in (18.59).

242

The variational equation for (18.61) is

$$(18.63) \qquad\qquad \lambda Cv = -z,$$

with the solution

$$(18.64) \qquad\qquad v = -C^{-1}z/\lambda.$$

Hence, the minimum value, obtained by direct substitution, is the quadratic form $-(C^{-1}z, z)/\lambda$.

It follows that (18.59) reduces to the form

$$(18.65) \qquad\qquad \frac{\partial f}{\partial a} = \sum_{i,j=1}^{N} d_{ij} \frac{\partial f}{\partial b_i} \frac{\partial f}{\partial b_j},$$

where the coefficients d_{ij} are known.

Using the fact that f has the form indicated in (18.58), we derive a system of ordinary quadratic differential equations of the form

$$(18.66) \qquad\qquad \frac{d}{da} q_{ij}(a) = \sum_{k,\ell=1}^{r} d_{k\ell} q_{ki}(a) q_{\ell j}(a),$$

with the initial conditions now at $a = T$, namely $f(b_1, b_2, \ldots, b_k; T) = g(b_1, b_2, \ldots, b_k)$. This relation determines the values of the functions $q_{ij}(a)$ at $a = T$.

Since the numerical solution of a system of one hundred nonlinear differential equations of this type is a completely routine matter with modern computers, we have a feasible algorithm, provided that r does not exceed a quantity of the order of magnitude of twenty or thirty. The dimension of the system in (18.66) is $r(r + 1)/2$.

Both methods require the computation of the matrix e^{At}. This is an initial value problem requiring the solution of N sets of N-dimensional linear differential equations.

Similar results can be obtained for stochastic and adaptive control processes. References will be found at the end of the chapter.

18.12 Successive Approximations Once More

These results open the path to the application of successive approximations in a variety of ways. Let us begin by considering the problem of minimizing the functional

$$(18.67) \quad J(y) = g[x_1(T), x_2(T), \ldots, x_N(T)] + \lambda \int_0^T h(y_1, y_2, \ldots, y_N)\, dt,$$

where x and y are related by means of the vector equation

$$(18.68) \qquad\qquad \frac{dx}{dt} = Ax + y, \quad x(0) = c.$$

243

Let $y^0 = [y_1^0, y_2^0, \ldots, y_N^0]$ be an initial guess in policy space, and let $x^0 = [x_1^0, x_2^0, \ldots, x_N^0]$ be the associated x_i-values, derived from (18.68).

As the next step, consider the problem of minimizing the functional

$$(18.69) \qquad J_1(y) = g[x_1(T), x_2^0(T), \ldots, x_N^0(T)] + \lambda \int_0^T h(y) \, dt,$$

subject to the relation of (18.68).

Since $J_1(y)$ depends only upon one component of $x(T)$, we can solve the problem computationally using sequences of functions of one variable.

Let

$$(18.70) \qquad x^1 = [x_1^1, x_2^1, \ldots, x_N^1], \quad y^1 = [y_1^1, \ldots, y_N^1]$$

be the set of x and y functions obtained in this way. We can now continue the approximation procedure in various ways. However, it seems very difficult to obtain monotone convergence.

18.13 Successive Approximations—II

Generally, the differential equation connecting x and y is nonlinear,

$$(18.71) \qquad \frac{dx}{dt} = g(x, y), \quad x(0) = c.$$

Using a linear approximation to this equation, say

$$(18.72) \qquad \frac{dx_{n+1}}{dt} = g_1(x_n, y_n) + Ax_{n+1} + By_{n+1},$$

obtained by expanding $g(x, y)$ around $x = x_n$, $y = y_n$, and retaining only the linear terms, we can combine the previous approximation methods with this.

18.14 Functional Approximation

Let us now discuss a totally different type of approximation technique with apparently unlimited possibilities not only in the domain of dynamic programming, but in the whole field of mathematical physics and generally wherever the computational solution of functional equations is important.

We have so far considered that a function $u(t)$ defined over $[0, 1]$, say, is to be represented by means of a table of values $\{u(k\delta)\}$, $k = 0, 1, \ldots, N$. The more accurate the representation, the finer the grid, and consequently the bigger the table of values. It is the size of the table required to store a function of more than two variables that makes us regard multidimensionality as a malediction.

To avoid these difficulties to a great extent, we take a more analytic viewpoint. Suppose that we approximate to $u(t)$ by means of a polynomial over $[0, 1]$,

$$(18.73) \qquad u(t) \cong a_0 + a_1 t + \ldots + a_k t^k.$$

The function $u(t)$ is now specified by $k + 1$ coefficients $a_0, a_1, \ldots, a_k$. Since one can obtain remarkably accurate approximations with polynomials of degree five or ten, we see that this method of storing a function $u(t)$ is quite efficient.

Turning to functions of two variables, $u(s, t)$, we see that an approximation such as

$$(18.74) \qquad u(s, t) \cong \sum_{i,j=0}^{k} a_{ij} s^i t^j,$$

requires $(k + 1)(k + 2)/2$ coefficients. Once again as compared to a table of values of a function of two variables, this represents an efficient technique.

In place of using a polynomial approximation, we can use a readily computed orthonormal sequence such as the Legendre polynomials, trigonometric functions, or Cebycev polynomials. Consider, as an indication of the method, the problem of computing the sequence $\{f_n(t)\}$ determined by the recurrence relation

$$(18.75) \qquad f_n(t) = \underset{0 \leq s \leq 1}{\text{Max}} \left[g(s, t) + f_{n-1}[h(s, t)] \right],$$

where $0 \leq h(s, t) \leq 1$ for all s and t in $[0, 1]$, and $f_0(t)$ is a given function.

At the n-th stage, the function $f_{n-1}(t)$ is determined by means of the sequence of coefficients $[a_0^{(n-1)}, a_1^{(n-1)}, \ldots, a_k^{(n-1)}]$ stored in the memory of the computer, and the formula

$$(18.76) \qquad f_{n-1}(t) = a_0^{(n-1)} + a_1^{(n-1)} \phi_1(t) + \ldots + a_k^{(n-1)} \phi_k(t),$$

where $\{\phi_i(t)\}$ is an orthonormal sequence of the type mentioned above. When $f_n(t)$ is determined from (18.75), it is used to obtain the sequence $[a_0^{(n)}, a_1^{(n)}, \ldots, a_k^{(n)}]$ by means of the formula

$$(18.77) \qquad a_i^{(n)} = \int_0^1 f_n(t) \phi_i(t) \, dt, \quad i = 1, 2, \ldots, k.$$

The question that now arises is that of evaluating the coefficient $a_i^{(n)}$. Using a quadrature formula, we have

$$(18.78) \qquad a_i^{(n)} = \sum_{j=1}^{M} c_j \phi_i(t_j) f_n(t_j).$$

It follows that it is necessary only to evaluate the values $f_n(t_j), j = 1, 2, \ldots, M$, at each stage.

It is clear that a computational scheme of this type will require a greater amount of time than those we have described in the previous chapters. It is, however, a technique that can be used to treat processes involving sequences of functions of three, four, or more variables. What we are doing is trading time, which we do possess, although expensive, for memory capacity, which exists only in limited quantities, regardless of expense.

18.15 Simulation Techniques

In many complex situations we may not have enough information, or too much information and insufficient resources or time, to formulate a decision process in quantitative terms. In some cases, particularly those involving human beings, we may not be able to develop precise cause and effect relations, or to develop a criterion function for evaluating the outcome of a sequence of decisions. Nevertheless, it is still sensible to talk about policies and approximations in policy space, even though the return functions may not exist and there may be no functional equations to analyze.

Using various forms of crude comparison, we can test different policies. In view of the vague evaluation which is performed, it is more than ever important to concentrate upon policies with simple, intuitive structure which are known to work well over a spectrum of processes. It is essential that we devise a method for examining the outcomes of these policies. One method for doing this is based upon the construction of a *simulation* process or *gaming* process.

Since any discussion, of even a half-hearted type, would take us too far afield, we refer the interested reader to the works cited at the end of the chapter. Many other references will be found in these.

18.16 Quasi-optimal Policies

Another way to handle processes in which no specific criterion function exists is to show that certain policies are optimal for a range of criterion functions.

As an example of how we are led to approximate policies of this nature, consider the following process. We have a resource, in quantity c, which may be utilized in whole or in part at each stage of a process. If y is utilized at a particular stage, there is an immediate return of $g(c, y)$, and a diminution in resources of $h(c, y)$. If, as usual, we denote by $f(c)$ the total return obtained using an optimal allocation policy, we obtain the equation

$$(18.79) \qquad f(c) = \operatorname*{Max}_{y} [g(c, y) + f\{c - h(c, y)\}].$$

Suppose that $h(c, y)$ is positive and small compared to c, and that $f'(c)$ is positive, a reasonable assumption. Then the foregoing equation takes the approximate form

$$(18.80) \qquad f(c) = \operatorname*{Max}_{y} [g(c, y) + f(c) - h(c, y)f'(c) + \ldots].$$

This, in turn, reduces to the result

$$(18.81) \qquad f'(c) \cong \operatorname*{Max}_{y} [g(c, y)/h(c, y)].$$

The most important consequence of this is not the expression for $f'(c)$, but rather the approximate optimal policy which it yields. Verbally, it expresses the idea that in a multistage process, the policy which maximizes the ratio of return to expenditure is close to optimal. This is a policy which can be used in a large number of cases.

It is not difficult to extend the foregoing analysis to obtain a similar result for stochastic allocation processes. We find that an approximate policy here is to maximize expected gain over expected cost.

18.17 Non-zero Sum Games

As mentioned in the chapter on game theory, there exists no satisfactory theory for non-zero sum games, and the evidence is rather overwhelming that there never will be.

Let us take a game of this nature, characterized by a matrix (a_{ij}) for A and a matrix (b_{ij}) for B, and change it into a game of survival in which A and B seek to ruin each other. Defining in the customary way

(18.82) $f(x, y) = $ the probability that A ruins B when A has x, B has y
and both employ optimal policies,

we obtain the functional equation

(18.83) $$f(x, y) = \underset{p}{\text{Max}} \underset{q}{\text{Min}} \left[\sum_{i,j} p_i q_j f(x + a_{ij}, y + b_{ij}) \right]$$
$$= \underset{q}{\text{Min}} \underset{p}{\text{Max}} \left[\qquad \cdots \qquad \right],$$

with the boundary conditions

(18.84) $$f(x, y) = 1, \quad x > 0, \quad y \le 0,$$
$$= 0, \quad x \le 0, \quad y > 0,$$
$$= \tfrac{1}{2}, \quad x = y = 0 \quad \text{(by convention)}.$$

Assume that a_{ij} and b_{ij} are small compared to x and y, so that we may write

(18.85) $$f(x + a_{ij}, y + b_{ij}) \cong f(x, y) + a_{ij}f_x + b_{ij}f_y.$$

Then (18.83) becomes

(18.86) $$f(x, y) \cong \underset{p}{\text{Max}} \underset{q}{\text{Min}} \left[\sum_{i,j} p_i q_j (f(x, y) + a_{ij}f_x + b_{ij}f_y) \right]$$
$$\cong \underset{q}{\text{Min}} \underset{p}{\text{Max}} \left[\qquad \cdots \qquad \right].$$

This, in turn, yields

(18.87) $$0 \cong \underset{p}{\text{Max}} \underset{q}{\text{Min}} \left[f_x \sum_{i,j} a_{ij} p_i q_j + f_y \sum_{i,j} b_{ij} p_i q_j \right]$$
$$\cong \underset{q}{\text{Min}} \underset{p}{\text{Max}} \left[\qquad \cdots \qquad \right].$$

247

Suppose now that $f_x > 0$, $f_y > 0$, and that one of the expected values, $\sum_{i,j} a_{ij} p_i q_j$ or $\sum_{i,j} b_{ij} p_i q_j$, say $\sum_{i,j} a_{ij} p_i q_j$, is of fixed sign for all p_i and q_j satisfying the conditions

(18.88)   (a) $p_i, q_j \geq 0$,

         (b) $\sum_i p_i = \sum_j q_j = 1$.

Then, (18.87) yields

(18.89) $$-\frac{f_x}{f_y} \cong \operatorname*{Max}_{p} \operatorname*{Min}_{q} \left[\sum_{i,j} b_{ij} p_i q_j \Big/ \sum_{i,j} a_{ij} p_i q_j \right]$$

$$\cong \operatorname*{Min}_{q} \operatorname*{Max}_{p} \left[\quad \cdots \quad \right].$$

The remarkable thing about this last equation is that the right-hand side is independent of the function $f(x, y)$. Since we obtain a similar approximate equation for a whole class of processes ruled by different criteria, but still possessing the same functional equation, it follows that the foregoing policy of min-maxing the quotient of the stagewise expected values is an approximate optimal policy for each of these.

Consequently, if we believe that a non-zero sum process is ruled by *some* criterion, known or unknown, then we can use the foregoing policy as an excellent approximation, under the preceding assumptions.

This situation is quite common in mathematical physics. We seek the properties of solutions of equations of precisely the same form, differing only in the boundary conditions, e.g. the shape of the region under consideration. In these cases, it is well known that far from the boundary the behavior of the solutions are very similar. This heuristic principle enables one to obtain a number of salient features of the solutions by comparison with the solutions of far simpler problems.

Were it not for simplifying ideas of this type, mathematical physics would be a far more difficult field than it is, and the unhappy physicist, clashing by night on a darkling plain, even unhappier.

18.18 Discussion

What we have tried to emphasize in the foregoing chapter is that the future of theories of multistage processes, decision or otherwise, lies in the direction of various sophisticated uses of the method of successive approximations. We may expect that as our understanding becomes more profound, our techniques will become more elegant and superficially simpler. The whole history of science teaches us, however, that this simplicity and depth of understanding are not easily won. Furthermore, they are won rather by example than by precept.

Bibliography and Discussion

18.1 The classical paper of Picard is

E. Picard, "Mémoire sur la Théorie des Equations . . .," *Jour. de Math.*, (4), 6(1890), pp. 145–210.

18.2 For a discussion of successive approximations applied to differential equations, see

E. Cotton, "Approximations successives et les equations differentiettes," *Mem. Sci. Fac.*, vol. XXVIII, Hermann et Cie., Paris, 1928.

R. Bellman, "On the boundedness of solutions of nonlinear differential and difference equations," *Trans. Amer. Math. Soc.*, vol. 46, 1948, pp. 354–388, where, in addition to several other methods, the Birkhoff-Kellogg fixed-point theorem is used, and

R. Bellman, *Stability Theory of Differential Equations*, McGraw-Hill Book Co., Inc., New York, 1954.

18.3 and 18.4 See

R. Bellman, *Dynamic Programming*, Princeton Univ. Press, Princeton, N.J., 1957.

18.5 See Chapter XII for a more detailed discussion.

18.7 See

R. Bellman, "Some new techniques in the dynamic programming solution of variational problems," *Q. Appl. Math.*, vol. 16, 1958, pp. 295–305, where a number of these ideas are discussed.

18.10 For the requisite background in differential-difference equations, see

R. Bellman and J. M. Danskin, *A Survey of the Mathematical Theory of Time-lag, Retarded Control, and Hereditary Processes*, The RAND Corporation, Report R-256, 1954,

and the forthcoming book,

R. Bellman and K. Cooke, *Differential-difference Equations*, Academic Press, New York, to appear.

18.11 This subject is discussed in great detail in the thesis,

R. Beckwith, *Analytic and Computational Aspects of Dynamic Programming Processes of High Dimension*, Ph.D. Thesis, Purdue University, June, 1959.

R. Bellman and R. Kalaba, *Reduction of Dimensionality, Dynamic, Programming and Control Processes*, The RAND Corporation, Paper P-1964, 1960.

For stochastic and adaptive processes, see

M. Freimer, *A Dynamic Programming Approach to Adaptive Control Processes*, Lincoln Lab. Report, 1959, Group Report 54-2.

D. Kramer, "On the Control of Linear Systems with Time Lags," P-1948, The RAND Corporation, 1960.

18.14 Some preliminary results of this type may be found in

R. Bellman and S. Dreyfus, "Functional Approximations and Dynamic Programming," *Math Tables . . .*, vol. XIII, 1959, pp. 247–251.

See also

C. W. Merriam, III, "An optimization theory for feedback control design," *Information and Control*, vol. 3, 1960, pp. 32–59.

The problem of determining the most efficient way of computing the value of a polynomial $p_k(t) = a_0 + a_1 t + \ldots + a_k t^k$ given the set of coefficients is one of great interest and difficulty. See

A. Ostrowski, *On Two Problems in Abstract Algebra Connected with Horner's Rule*, Studies in Mathematics and Mechanics, Academic Press, 1954, pp. 40–48.

The general question of approximation by polynomials and rational functions is taken up in

C. Hastings, *Approximations for Digital Computers*, Princeton Univ. Press, Princeton, N.J., 1955.

Finally, let us mention a new technique due to G. Kron for decomposing complex processes into simple, the "tearing method." See

J. P. Roth, "An application of algebraic topology, Kron's method of tearing," *Q. Appl. Math.*, vol. XVII, 1959, pp. 1–23.

J. P. Roth, "An application of algebraic topology: Kron's method of tearing," *Q. Appl. Math.*, vol. 17, 1959, pp. 1–24.

INDEX

PUBLISHED RAND RESEARCH

PRINCETON UNIVERSITY PRESS, PRINCETON, NEW JERSEY

Approximations for Digital Computers, by Cecil Hastings, Jr., 1955
International Communication and Political Opinion: A Guide to the Literature, by Bruce Lannes Smith and Chitra M. Smith, 1956
Dynamic Programming, by Richard Bellman, 1957
The Berlin Blockade: A Study in Cold War Politics, by W. Phillips Davison, 1958
The French Economy and the State, by Warren C. Baum, 1958
Strategy in the Missile Age, by Bernard Brodie, 1959
Foreign Aid: Theory and Practice in Southern Asia, by Charles Wolf, Jr., 1960

COLUMBIA UNIVERSITY PRESS, NEW YORK, NEW YORK

Soviet National Income and Product, 1940–48, by Abram Bergson and Hans Heymann, Jr., 1954
Soviet National Income and Product in 1928, by Oleg Hoeffding, 1954
Labor Productivity in Soviet and American Industry, by Walter Galenson, 1955

THE FREE PRESS, GLENCOE, ILLINOIS

Psychosis and Civilization, by Herbert Goldhamer and Andrew W. Marshall, 1953
Soviet Military Doctrine, by Raymond L. Garthoff, 1953
A Study of Bolshevism, by Nathan Leites, 1953
Ritual of Liquidation: The Case of the Moscow Trials, by Nathan Leites and Elsa Bernaut, 1954
Two Studies in Soviet Controls: Communism and the Russian Peasant, and Moscow in Crisis, Herbert S. Dinerstein and Leon Gouré, 1955
A Million Random Digits with 100,000 Normal Deviates, by The RAND Corporation, 1955

HARVARD UNIVERSITY PRESS, CAMBRIDGE, MASSACHUSETTS

Smolensk Under Soviet Rule, by Merle Fainsod, 1958
The Economics of Defense in the Nuclear Age, by Charles J. Hitch and Roland McKean, 1960

MCGRAW-HILL BOOK COMPANY, INC., NEW YORK, NEW YORK

The Operational Code of the Politburo, by Nathan Leites, 1951
Air War and Emotional Stress: Psychological Studies of Bombing and Civilian Defense, by Irving L. Janis, 1951
Soviet Attitudes Toward Authority: An Interdisciplinary Approach to Problems of Soviet Character, by Margaret Mead, 1951
Mobilizing Resources for War: The Economic Alternatives, by Tibor Scitovsky, Edward Shaw, and Lorie Tarshis, 1951
The Organizational Weapon: A Study of Bolshevik Strategy and Tactics, by Philip Selznick, 1952
Introduction to the Theory of Games, by J. C. C. McKinsey, 1952
Weight-Strength Analysis of Aircraft Structures, by F. R. Shanley, 1952
The Compleat Strategyst: Being a Primer on the Theory of Games of Strategy, by J. D. Williams, 1954
Linear Programming and Economic Analysis, by Robert Dorfman, Paul A. Samuelson, and Robert M. Solow, 1958
Introduction to Matrix Analysis, by Richard Bellman, 1960
The Theory of Linear Economic Models, by David Gale, 1960

THE MICORCARD FOUNDATION, MADISON, WISCONSIN

The First Six Million Prime Numbers, by C. L. Baker and F. J. Gruenberger, 1959

NORTH-HOLLAND PUBLISHING COMPANY, AMSTERDAM, HOLLAND

A Time Series Analysis of Interindustry Demands, by Kenneth J. Arrow and Marvin Hoffenberg, 1959

FREDERICK A. PRAEGER, PUBLISHERS, NEW YORK, NEW YORK

War and the Soviet Union: Nuclear Weapons and the Revolution in Soviet Military and Political Thinking, by H. S. Dinerstein, 1959

PUBLIC AFFAIRS PRESS, WASHINGTON, D.C.

The Rise of Khrushchev, by Myron Rush, 1958
Behind the Sputniks: A Survey of Soviet Space Science, by F. J. Krieger, 1958

RANDOM HOUSE, INC., NEW YORK, NEW YORK

Space Handbook: Astronautics and Its Applications, by Robert W. Buchheim and the Staff of the RAND Corporation, 1959

ROW, PETERSON AND COMPANY, EVANSTON, ILLINOIS

German Rearmament and Atomic War: The Views of German Military and Political Leaders, by Hans Speier, 1957
West German Leadership and Foreign Policy, edited by Hans Speier and W. Phillips Davison, 1957
The House Without Windows: France Selects a President, by Constantin Melnik and Nathan Leites, 1958
Propaganda Analysis: A Study of Inferences Made from Nazi Propaganda in World War II, by Alexander L. George, 1959

STANFORD UNIVERSITY PRESS, STANFORD, CALIFORNIA

Strategic Surrender: The Politics of Victory and Defeat, by Paul Kecskemeti, 1958
On the Game of Politics in France, by Nathan Leites, 1959
Atomic Energy in the Soviet Union, by Arnold Kramish, 1959
Marxism in Southeast Asia: A Study of Four Countries, edited by Frank N. Trager, 1959

JOHN WILEY & SONS, INCORPORATED, NEW YORK, NEW YORK

Efficiency in Government through Systems Analysis: With Emphasis on Water Resource Development, by Roland N. McKean, 1958